MATERIALS SCIENCE AND TECHNOLOGIES

ADHESIVE PROPERTIES IN NANOMATERIALS, COMPOSITES AND FILMS

Materials Science and Technologies

Additional books in this series can be found on Nova's website
under the Series tab.

Additional E-books in this series can be found on Nova's website
under the E-books tab.

MATERIALS SCIENCE AND TECHNOLOGIES

ADHESIVE PROPERTIES IN NANOMATERIALS, COMPOSITES AND FILMS

KERI A. WILKINSON
AND
DANIEL A. ORDONEZ
EDITORS

Nova Science Publishers, Inc.
New York

For permission to use material from this book please contact us:
Telephone 631-231-7269; Fax 631-231-8175
Web Site: http://www.novapublishers.com

Library of Congress Cataloging-in-Publication Data

Adhesive properties in nanomaterials, composites, and films / editors, Keri
A. Wilkinson and Daniel A. Ordonez.
p. cm.
Includes bibliographical references and index.
ISBN 978-1-61209-268-3 (hardcover)
1. Adhesives. I. Wilkinson, Kerry. II. Ordonez, Daniel A.
TA455.A34A34 2010
620.1'99--dc22
2010048339

Published by Nova Science Publishers, Inc. † New York

Contents

PREFACE

This new book examines the adhesive properties in nanomaterials, composites and films. Topics discussed include the properties and applications of composite materials made of Norland Optical Adhesive and liquid crystal materials; adhesive bonding of hydro-thermally modified wood; the use of adhesive films in transdermal and mucoadhesive dosage forms and the durability of adhesives and matrices for polymer composites used in the restoration and rehabilitation of building structures under natural and accelerated weathering conditions.

Chapter 1 - This chapter reviews properties and applications of composite materials made of Norland Optical Adhesive and liquid crystal materials. The polymer support exhibits the texture of a film with Swiss cheese morphology and microscopic inclusions filled with small functional liquid crystal molecules. Under certain conditions of preparation, these systems have a grating morphology with a succession of liquid crystal and polymer shells. Systems with Swiss cheese morphology and mean diameter of inclusions in the micrometer range are strong scatterers of visible light and their applications are mainly in technologies using commutable windows and display devices. Other applications are also possible in multiplexing devices and routing of telecommunication signals provided that the nature of constituents in the initial mixture and the method of preparation are properly chosen. In general, one seeks liquid crystal domains with sizes in the nanometer range either with randomly distributed inclusions or using grating morphology. This study presents a state of the art of the properties of such composite materials encompassing morphology, phase behavior, thermo-mechanical and visco-elasticity aspects. Optical and electro-optical responses which are the basis for numerous applications are also examined. Both theoretical models and experimental investigations will be considered with a special reference to the work with which the authors are most familiar. A correlative analysis in made to demonstrate the relationship between composition of the initial mixture, conditions and methods of preparation and final properties of the materials.

Chapter 2 - Bonded joints made with polymeric matrix composites are commonly adopted in structural applications where weight is a critical design parameter. They are also key elements in the repair and retrofitting of damaged structures, e.g. aircraft composite skin and reinforced concrete bridge columns. Advances in the design and inspection of bonded polymeric composite joints will therefore improve joints durability, and consequently the safety of composite structures, in a wide range of applications (aerospace, civil, ship, transportation and wind power engineering).

The scope of this chapter is to discuss one aspect of joint durability: chemical and mechanical degradation of the individual components of a typical aerospace joint, i.e.

structural epoxy-based adhesive and carbon/epoxy. In a recent research project, these materials were separately exposed to an aggressive environment, consisting of full immersion in water or anti-icing additive (also called fuel additive) or jet fuel or hydraulic fluid. There were simplified laboratory testing conditions: no coatings, no mixing of fluids (i.e. jet fuel and anti-icing additive), no prior thermo-mechanical damage. Gravimetric data, hardness tests and microscopy support the presence of chemical degradation in the adhesive. The use of simple Fickian and non-Fickian two-stage sorption Langmuir models for gravimetric data appears successful for the results of some treatments, e.g. sorption of fuel additive by adhesive. This finding could be used for the purpose of multiphysics modeling of thermo-mechanical degradation of bonded joints. Finally, chemical degradation distinctly appears through Differential Scanning Calorimetry (DSC) and thermogravimetric (TGA) tests: significant changes were encountered when the adhesive was treated in anti-icing additive or hydraulic fluid, while other treatments seem to be much less detrimental for the adhesive. Carbon/epoxy, on the other hand, is impacted at a much lesser rate by fuel additive.

Chapter 3 - The coating of polymer materials by protecting and functional films requires high efforts in the development of coating techniques due to the very different mechanical and thermal properties of polymer substrates and metal or ceramic films. The film has to fulfil both high adhesion and optimized microstructure to prevent failing in its application. The current work describes a new vacuum coating technique for polymer materials with inorganic films – the Pulsed Laser Deposition (PLD) process, characterized by a high-energetic pulsed plasma and the easy possibility to room temperature deposition (RT-PLD). Thus, pseudodiffusion interfaces were found due to the high-energetic particle bombardment during PLD coating. Additionally, changes of the polyurethane chemical binding are evident, like the transition from C=O to C–O–R binding, in which titanium atoms could act as new binding partners to the O species. Although very high film adhesion can be guaranteed in the PLD by the formation of pseudodiffusion interfaces, preventing the well-known buckling phenomenon, high film stresses result in plastic deformation of the soft polymer surface and the formation of wrinkles. The reasons and effects of wrinkling – even starting in growing films – on the film behaviour are described in this work, based on both practical investigations, using transmission electron microscopy, X-ray diffraction and atomic force microscopy, and theoretical finite element modelling.

Chapter 4 - In the past years considerable increase in the hydrothermal modification of wood was observed. Mostly the heat treatments are performed to change the hygroscopicity of wood. Furthermore, densification processes are utilizing the hydrothermal treatments. A key factor in the efficient utilization of timber resources is the adhesive bonding of wood, since manufacturing of wood based composites depends on forming bonds between individual wooden elements. Wood-based composites offer several advantages over sawn wood, such as the utilization of waste material, better distribution of non-homogeneities, and control of the product properties in the manufacturing process. Therefore the efficient utilization of hydro-thermally modified wood depends on its adhesive potential. The combined effects of temperature and moisture modify the properties of the polymeric components of wood and its porous structure. Wood tissue is exposed to high temperatures that can cause surface inactivation. Hydrothermal treatment could reduce the surface free energy and thus result in the poorer wettability of the modified wood surface. Furthermore, penetration and spreading of the resin could be influenced by hydrothermal treatment. In spite of numerous studies of hydro-thermally modified wood, the adhesion potential of hydrothermal treated wood has not been studied extensively in the past. The aim of this chapter is to provide literature review of aspects like surface properties of hydro-thermally modified wood related to bondability, wetting, and penetration. Finally, future directions regarding efficient application of hydro-thermally modified wood including densified wood in polymer composites are discussed.

Chapter 5 - Thin polymeric films with adhesive ability are useful for transdermal and mucoadhesive drug delivery systems. Polymer materials with adhesive ability in their dry state are integral to the formulation of patch systems for topical and transdermal drug delivery. Such polymers are often called "pressure sensitive adhesives" due to their capacity to attach to the skin surface with the application of light pressure. In the drug-in-adhesive design the drug is mixed with the adhesive polymer to produce a thin medicated film. The adhesive performance of these films can be monitored directly using tack and peel tests and indirectly by correlation with rheological parameters. Polymers with adhesive ability following absorption of moisture are useful in the formulation of mucoadhesive films for transmucosal (*e.g.* buccal, nasal, ocular) drug delivery. Such polymers are hydrophilic (hydrogels). Following hydration, polymer chains relax and interact with mucus glucoprotein chains, primarily by hydrogen bonding. This chapter will describe the properties of the adhesive polymers used in the design of transdermal and mucoadhesive films, the mechanism of adhesion and the tests that can be applied to monitor the adhesive performance.

Chapter 6 - In this chapter, a review of theoretical and numerical asymptotic studies on thin adhesive layers is proposed. A general mathematical method is presented for modelling the mechanical behavior of bonding and interfaces. This method is based on a simple idea that the adhesive film is supposed to be very thin; the mechanical problem depends strongly on the thinness of the adhesive. It is quite natural, mathematically and mechanically, to consider the limit problem, that is, the asymptotic problem obtained when the thickness and, possibly, the mechanical characteristics of the adhesive thin layer tend to zero. This asymptotic analysis leads to a limit problem with a mechanical constraint on the surface, to which the layer shrinks. The formulation of the limit problem includes the mechanical and geometrical properties of the layer. This limit problem is usually easier to solve numerically by using finite elements software. Theoretical results (i.e. limit problems) can be usually obtained by using at least four mathematical techniques: gamma-convergence, variational analysis, asymptotic expansions and numerical studies. In the chapter, some examples will be presented: comparable rigidity between the adhesive and the adherents, soft interfaces, adhesive governed by a non convex energy and imperfect adhesion between adhesive and adherents. Some numerical examples will also be given and, finally, an example of a numerical algorithm will be presented.

Chapter 7 - The success of the fiber-reinforced polymer (FRP) systems in the restoration and rehabilitation of civil and monumental structures is due to their excellent properties, generally superior than those of traditional building materials. Of great importance, however, is the behavior of the repaired structure under loading and its durability in the outside climate. The lack of specific standards for durability investigation of materials employed in such applications makes difficult the assessment of reliable theoretical models. As an example, the available standard tests for adhesives generally refer to resins cured at elevated temperatures, neglecting the peculiarities of "cold-cured" adhesives.

In this chapter, the durability of the base components of FRP specifically designed for civil engineering industry, is reviewed. The most common environmental agents, mostly responsible for the deterioration of the materials performance, are examined. Finally, standardized accelerating tests are discussed as an effective method to predict the long term behavior of the weathered materials.

Chapter 8 - Adhesives or bonding agents surround all living beings in Nature and in their daily lives. Adhesives or bonding agents are used in a variety of industries: construction, packaging, furniture, automotive, appliance, textile, aircraft, and many others. However, most of them are toxic for human beings due to the presence of harmful synthetic additives. The long-term exposure to these toxic substances can cause a range of ailments, including cancer, asthma, and Alzheimer disease. The ultimate solution of this trend lies in the emulation of Nature following the true pathway of natural process. In this paper, a number of adhesives have been formulated from natural additives, which can lead us towards sustainable lifestyles. Experiments show that the product exhibit durability and strength comparable to commercially available products, while the required concentration of the adhesive is low. Further experiments indicate that alternative adhesives that are of organic origin can eliminate the use of synthetic adhesives entirely. Similarly, as a natural alternative to Plaster of Paris is proposed. This product has features very similar to the commercial Plaster of Paris. This product and others can become useful for people with chemical sensitivity, particularly the children and the elderly. This paper describes in detail the specific recommendations on natural adhesives, their strengths and durability as a function of temperature.

In: Adhesive Properties in Nanomaterials, Composites ... ISBN: 978-1-61209-268-3
Editors: K. A. Wilkinson and D. A. Ordonez

Chapter 1

NORLAND OPTICAL ADHESIVE AND LIQUID CRYSTAL COMPOSITE MATERIALS

Réda Benmouna*[*] *and Mustapha Benmouna
University Aboubekr Belkaïd, Faculty of Sciences,
Department of Physics, Tlemcen BP119, Algeria.

ABSTRACT

This chapter reviews properties and applications of composite materials made of Norland Optical Adhesive and liquid crystal materials. The polymer support exhibits the texture of a film with Swiss cheese morphology and microscopic inclusions filled with small functional liquid crystal molecules. Under certain conditions of preparation, these systems have a grating morphology with a succession of liquid crystal and polymer shells. Systems with Swiss cheese morphology and mean diameter of inclusions in the micrometer range are strong scatterers of visible light and their applications are mainly in technologies using commutable windows and display devices. Other applications are also possible in multiplexing devices and routing of telecommunication signals provided that the nature of constituents in the initial mixture and the method of preparation are properly chosen. In general, one seeks liquid crystal domains with sizes in the nanometer range either with randomly distributed inclusions or using grating morphology. This study presents a state of the art of the properties of such composite materials encompassing morphology, phase behavior, thermo-mechanical and visco-elasticity aspects. Optical and electro-optical responses which are the basis for numerous applications are also examined. Both theoretical models and experimental investigations will be considered with a special reference to the work with which the authors are most familiar. A correlative analysis in made to demonstrate the relationship between composition of the initial mixture, conditions and methods of preparation and final properties of the materials.

---

[*] E-mail: redabenmouna@yahoo.com

1. Introduction

Norland Optical Adhesive in short NOA is a commercial product made essentially of a mixture of mercapto-ester with acrylate monomer. Its precise composition is not disclosed in the open literature but with the information available, it is possible to construct its molecular structure. It contains in addition to benzophenone, the 3 types of monomers: trimethylolpropane diallyl ether, trimethylolpropane tristhiol and isophorone di isocyanate ester. [1,2] The photo-initiator benzophenone which is present at a known concentration is very efficient and has a high absorption cross section in the UV-visible and IR spectrum. Different types of adhesives may be encountered depending on composition and dedicated applications. NOA65 is among the most used ones in fundamental studies. It is an efficient non toxic adhesive with a capacity to form densely cross-linked networks under UV curing. Thiol radicals present in NOA65 contain sulfur groups (-SH) that allow polymerization under ordinary atmospheric environment without the inhibition by O_2 often encountered in the chemistry of unsaturated polyesters and acrylates. Its curing kinetics is much less sensitive to material mass than other hardeners and can be used in thin film cures.

Cure time depends upon dose, intensity and wavelength. The radicals initiate a cascade of polymerization reactions in the presence of monomers with double bonded vinyl groups. At a certain time, a phase separation mechanism takes place. While the polymerization is initiated by production of free radicals via radiation exposure, the phase separation mechanism is governed by combined changes in the system composition and its solubility properties. At final stages of polymerization, the system transforms into a more or less hard solid consisting of a cross-linked polymer network.

UV photo-polymerization of NOA is used in a variety of high performance coating for vinyl or wood flooring, metal and wood furniture, printing plastics, paper packaging, circuit board coating, and metal containers.[3-5]

Mixing such a solid material with a low molecular weight liquid crystal yields thin films that can be used as a support in photonic devices, microelectronic assemblies and hybrid organic – inorganic compounds. Such technologies are nowadays under constant development for many applications including display and telecommunications devices.[6-8]

In view of their importance both from the fundamental and applications standpoints, it would be useful to examine the properties of these compounds in order to advance our understanding on their specific behavior and elucidate some of their intriguing properties. The present work is a contribution along these lines scrutinizing different properties directly related with applications.

This chapter is organized as follows. In the first section we discuss the method of film preparation and morphology development. A strong correlation exists between composition of initial mixture, conditions of preparation and properties of the synthesized product. A special focus is put on the photo-curing process via Polymerization Induced Phase Separation (in short PIPS). The fundamental properties are intimately related with the final morphology and morphology development in the course of film preparation. The type of application is often conditioned by morphology considerations. The phase behavior, thermo-physical, thermo-mechanical and visco-elastic properties are the subject of sections 2 and 3. Evolution of the phase diagram from the initial state to the final cured film retains a particular attention. Transition temperatures of major events characterizing the system may be examined using a variety of techniques. The mechanical tenure and visco-elasticity of films subject to repetitive use under extreme conditions bear a strong impact on their performances and ageing and should be analyzed carefully. These aspects are also examined. Questions related with optical and electro-optical properties are at the basis of most applications in a variety of domains leading to new routes of development in modern technologies. These are the subject of section 4. Conclusion gives a brief synthesis of results.

2. Polymerization Induced Phase Separation and Morphology Development

2.1. Polymerization Induced Phase Separation

There are at least two main routes for fabricating polymer and liquid crystal composites materials. One consists of mixing together the pre-polymer and liquid crystal in the presence of a compatibilizer. This is the common way taken to obtain polymer dispersed liquid crystal films for privacy windows at the industrial level. Since the polymer and liquid crystal are not miscible under normal atmospheric conditions the compatibilizer which consists of water and an emulsifier (Sodium Dodecyl Sulfate in short SDS) is added to obtain a miscible system. After reaching the equilibrium state of a completely mixed system, the solvent is evaporated to obtain a film with Swiss cheese morphology whereby liquid crystal molecules are confined in the polymer pores. [9,10] The second route is the one that retain our attention in this chapter since it is commonly used to investigate the fundamental aspects of such multi-component mixtures. It consists of adding monomer and liquid crystal prior to polymerization and is referred to as Polymerization Induced Phase Separation. It leads to a variety of interesting properties for fundamental studies as we shall demonstrate throughout different parts of this chapter.[11,12]

Interplay between kinetics of polymerization and phase separation in polymer and liquid crystal materials is dominated primarily by the increasing molar weight and gelation mechanism of the polymer network. Elastic forces at the polymer and liquid crystal interface play a key role in the morphology development and performances of the film. [13]

The phase separation mechanism starts with nucleation of liquid crystal domains driven by monomer reaction kinetics. As this process evolves, nematic domains grow via Oswald ripening and diffusion mechanisms until completion of monomer consumption. [14] The final state exhibits a characteristic swiss cheese morphology with pores filled with small liquid crystal molecules distributed throughout the polymer network.

A low curing intensity leads to slow kinetics of polymerization and large liquid crystal domains. The lower is the rate of polymerization, the larger the size of these domains. Under such conditions, the composite is characterized by a high level of scattering in the visible range which is perfectly suitable for display devices. [15,16] Conversely, using high intensity curing sources speeds up the polymerization kinetics shifting the size to the submicron or even the nanometer range. At a certain stage of the polymerization kinetics, solvent molecules may exhibit a poor solvatation with respect to the growing polymer while monomer exhibits a higher miscibility, then nuclei coalesce to form large aggregates separated by inclusions filled with liquid crystal molecules. If the polymerization rate is fast, the transition to a network polymer is accelerated leading to nano sized droplets distributed throughout the volume. On the other hand, a high intensity source may lead to poor phase separated not uniformly distributed inclusions with diameters ranging from say 200 to 650nm. Low intensity sources yield nearly spherical domains in the micrometer range, well distributed throughout the volume. Small inclusions are found near the surface directly exposed to the source. [17] During curing the temperature rises faster for higher source intensity. The rate of nucleation and growth of liquid crystal domains depends on activation energies of nucleation and diffusion. There is a subtle interplay between these activation processes especially in terms of temperature and radiation dose. Viscosity of the medium plays a determinant role in favoring one or the other process.[18]

From the theoretical point of view, one has to solve coupled equations for the polymerization kinetics and dynamics of phase separation. To this end, Cahn-Hilliard-Cook theory of spinodal decomposition is often combined with other theories such as Flory-Huggins lattice model to develop a formalism of phase separation and kinetics of polymerization. These models involve parameters related with the mobility of molecules within the systems together with interfacial parameters characterizing the degree of incompatibility between unlike species.[19-22] Important quantities such as diffusion coefficients and rate constants of polymerization should be properly incorporated in the model to solve the problem.[23] Note that curing radiation and polymerization induce substantial heating within and beyond the exposed region because of diffusion mechanism and this should be taken into consideration in modeling these processes.[24, 25]

2.2. Morphology Development

Morphology is the prerequisite to any investigation because it determines to a large extent the performance under practical conditions of applications. It is intimately related with the nature and composition of the initial mixture as well as with the conditions of preparation. The size of inclusions is within μm or nm and keeps increasing with the amount of liquid crystal. Similar trends are also found by fixing the latter quantity and lowering the temperature below the isotropic to nematic transition. Competing processes of polymerization and phase separation and their respective kinetics can be monitored to obtain one or the other of the following composites: Phase separated composites films [26], PDLCs [27] and Holographic PDLCs [28] are shown in figures 1 a, b and c, respectively.

In the first case, curing radiation penetrates the sample with an intensity gradient and the exposed side undergoes fast polymerization with significant space and time modulations. This leads to two distinct domains as shown in panel (a). The polymer film is formed on the exposed side while the back side which receives a lower intensity is mainly made of liquid crystal molecules. Polymer Dispersed Liquid Crystal (Panel b) consists of a Swiss cheese morphology whereby the liquid crystal phase in dispersed in the polymer matrix in the form of inclusions. Depending on the preparation conditions, in the course of polymerization and phase separation processes, a variety of mesoscopic structures emerge. The growing molecular weight of network generates elastic stresses that bear strong effects on morphology developments. Diffusion and coalescence phenomena, nematic, smectic or higher order structures have to be taken into account to describe these developments characterizing the phase separated composite material. [29] Optical microscopy is sufficient to resolve particle size distribution in the micron range whereas for nano structured systems a more appropriate observation technique should be considered based on electron or atomic force microscopy. In this case, a careful preparation of samples is necessary to enhance the contrast conditions together with powerful software to analyze size distributions. [30, 31]

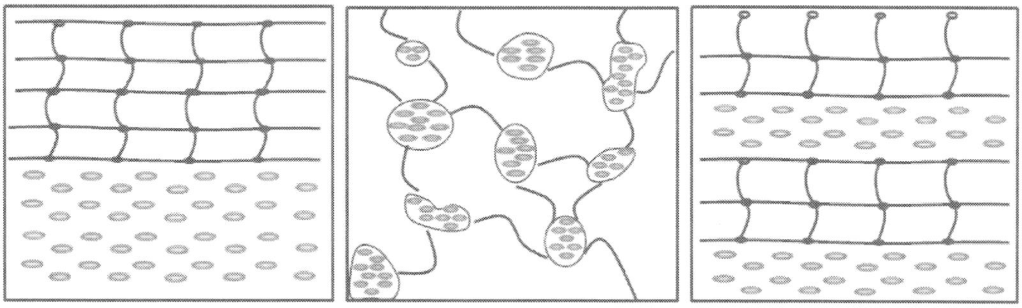

Figure 1. Schematic representation of (a) Phase Separated Composite films (b) Polymer Dispersed Liquid Crystals (c) Holographic Polymer Dispersed Liquid Crystal.

Specific morphologies are chosen according to the needs in terms of practical applications. A major effort is currently put on polymer dispersed liquid crystals with grating. These systems consist of arrays of liquid crystal and polymer thin layers with nano dimensions. [31] These arrays are built by implementing a procedure of preparation based on interferences between two laser beams. Such approach is suitable for systems like fiber optics, telecommunication routing and multiplexing, spectral selectivity and light valves with multiple wavelengths. [32] Photo-polymerization Induced Phase Separation mechanism may be used to achieve systems with random dispersions of nano-sized spherical domains filled with small liquid crystal molecules. These are characterized by an orientation order along a director axis. Orientation distribution throughout the polymer matrix is random in the quiescent state but may change when activated by an external field. These systems offer wide prospects in terms of applications comparable to systems exhibiting grating morphology of nano domains but their fabrication is much simpler. [33, 34]

Panel (a) of figure 2 shows the morphology developments upon cooling of the uncured NOE65/E7 system at a concentration of 70wt%E7. A transition from a single phase isotropic liquid to a biphasic system exhibiting a nematic structure is clearly seen. Another path was envisioned as illustrated in panel (b) of this figure where temperature was fixed and content of liquid crystal increased. On the left hand side, one observes a single phase of an isotropic liquid. When the amount of E7 increased, a nematic phase emerged with an increasing number of domains as one enters deeply into the biphasic region.

Figure 3 a, b, c show typical images obtained by Scanning Electron Microscope (SEM) of NOA65/E7 cured systems. These images are accompanied by histograms describing the pore size distribution within the polymer matrix. These are typical examples of morphology developments. Detailed investigations of pertinent parameters affecting the morphology of thiolene systems like these were reported elsewhere. [33-36] As the concentration of liquid crystal increases, pore size increases approaching the micrometer range. Porosity of the medium can be roughly estimated by the concentration of porogen which, in the present case is the low molecular weight liquid crystal.

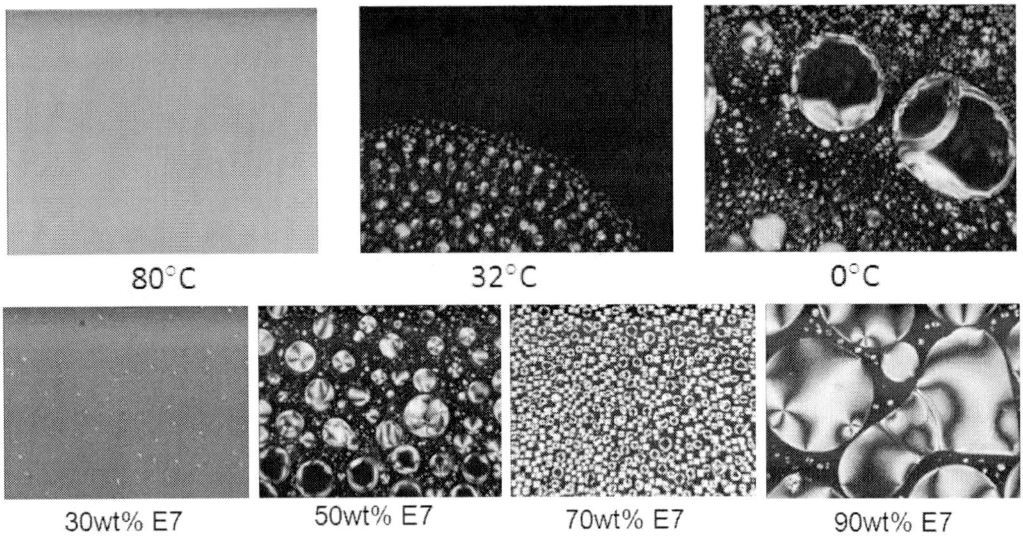

Figure 2. Morphology development of the uncured NOA65/E7 as obtained by Polarized Optical Microscopy (a) Upon cooling the mixture with 70wt% E7 (The corresponding temperature are indicated) (b) Increasing the content of E7 at 20°C (The concentration of E7 are specified on the figure).

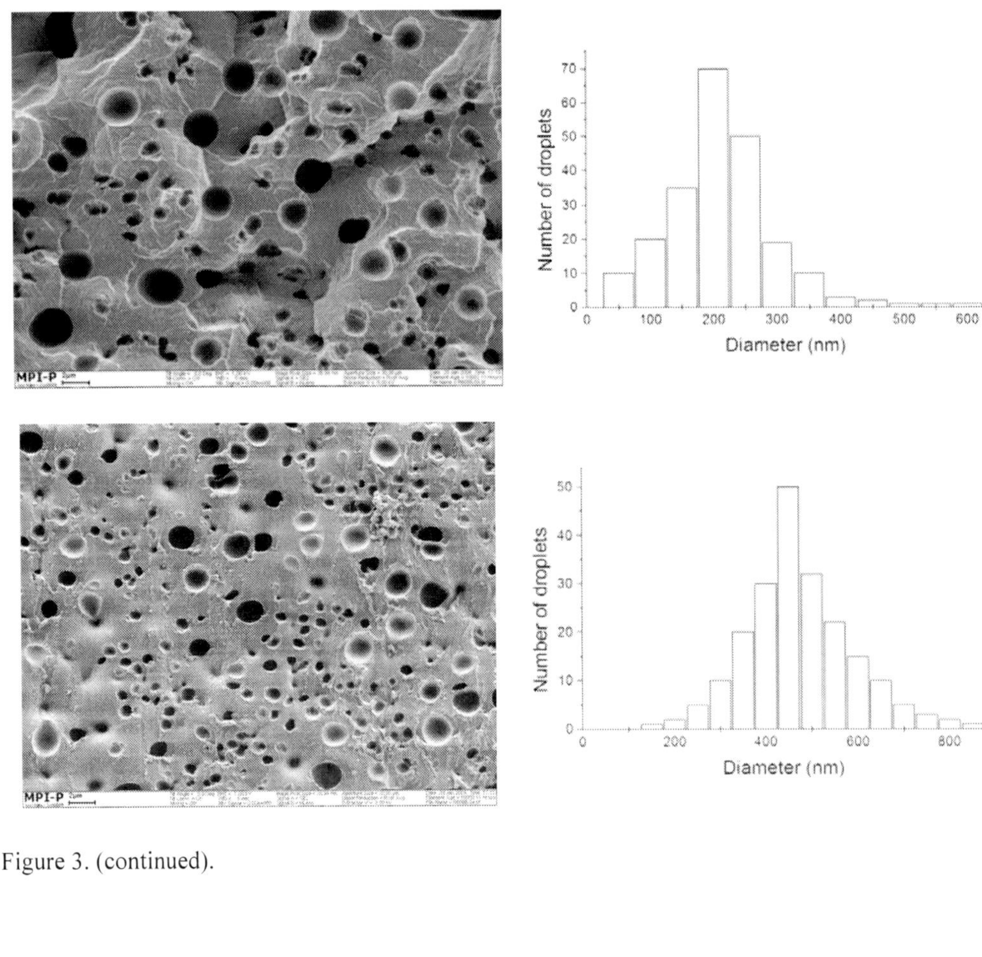

Figure 3. (continued).

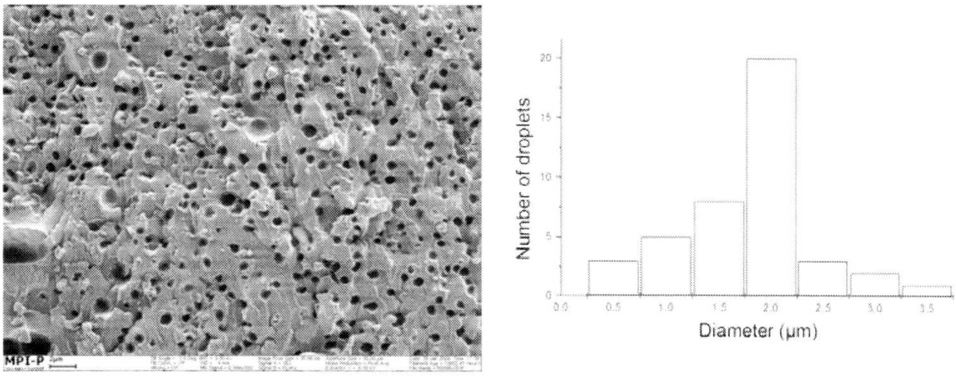

Figure 3. Morphology development of the cured NOA65/E7 system as obtained by Scanning Electron Microscopy (a) Micrograph and corresponding histogram at 35wt% E7 (b) 30wt% E7 (c) 25wt% E7.

A large increase in the pore size is not particularly suitable because of the scattering enhancement of visible light which could alter system efficiency. Conditions of photo-polymerization should be controlled with care since the dose of curing radiation, the nature and concentration of initiator and cross-linker should be chosen carefully to achieve the right system for the desired application. Higher functionality or concentration of cross-linker yield densely cross-linked polymer networks promoting the formation of smaller pores with the risks of generating heterogeneities.

For multiplexing devices, one would need a system with randomly dispersed nano-pores where the low molecular weight molecules remain totally confined. The amount of liquid crystal dispersed in the polymer matrix should be reduced for several reasons. Practical uses of these devices articulate around the ability of the liquid crystal molecules to respond properly and promptly to signal commands. Optimal conditions of the sensing behavior are achieved if all the LC molecules are fully confined in the pores. A way to control this situation is to measure the enthalpy exchange at the nematic to isotropic transition which, according to the method of Smith [37], gives the amount of molecules confined in pores. Small molecules dispersed in the supporting medium act as a plasticizer lowering its T_g with negative consequences on mechanical strength, ageing and performances.

Admitting that porosity is given by the porogen concentration, namely $\text{Porosity} = \frac{V_{pores}}{V_{total}} = \frac{\pi}{6}\frac{D^3(\text{number pores})}{V_{total}} = \phi_{LC}$, ones finds that the mean diameter of pores should scale like $\phi_{LC}^{1/3}$ which is in discrepancy with the experimental finding of references [33, 34] where D was found to be proportional to ϕ_{LC}. Such difference could be attributed to a lack of sufficient experimental data since only 3 measurements were reported and polydispersity may hide the real power law between ϕ_{LC} and D besides the fact that porosity should take into account the unavoidable amount of molecules that remain in the polymer matrix.

Obviously, the properties of these materials are highly sensitive to the quality of initial uncured systems. Aimed properties and applications correlate directly to the initial composition and temperature. This is why the phase diagram in the temperature composition frame is of primary importance to push for further investigations of the properties of polymer and liquid crystal composites.

3. Phase Behavior and Thermophysical Properties

3.1. Phase Behavior

The experimental phase behavior of low molecular weight liquid crystal and polymer systems can be rationalized using simple mean field theories combining the Flory-Huggins [38] lattice model for isotropic mixing and the Maier-Saupe [39] model for nematic order. This combination is appropriate to the initial uncured solution and to cured films involving linear polymers. Therefore, it would be useful to recall the main steps for the derivation of the phase diagram starting from the free energy. This can be written as a sum of the isotropic Flory-Huggins free energy density and the nematic Maier-Saupe contribution. We have

$$\frac{f_{FH}}{k_B T} = \frac{\phi_1}{N_1}\ln\phi_1 + \frac{\phi_2}{N_2}\ln\phi_2 + \chi\phi_1\phi_2 \tag{1}$$

where ϕ_1 and ϕ_2 are volume fractions of liquid crystal and monomer, N_1 and N_2 their respective numbers of repeat units and χ is the Flory interaction parameter. We shall assume in the case of uncured systems that $N_1 = N_2 = 1$ together with the incompressibility condition $\phi_1 + \phi_2 = 1$. The interaction parameter varies with temperature according to the common form $\chi = A + \frac{B}{T}$ where A and B are constants independent of temperature.

Orientation nematic order is described by the Maier-Saupe free energy expressed in term of the order parameter S and the partition function Z.

$$\frac{g_{MS}}{k_B T} = \frac{1}{N_1}\left[-\phi_1 \ln Z + \frac{1}{2}\nu\phi_1^2 S^2\right] \tag{2}$$

The order parameter is given by

$$S = \frac{1}{2}\left[3\left\langle\cos^2\theta\right\rangle - 1\right] \tag{3}$$

where θ is the angle between a reference axis and the director while the symbol <...> represents an average with respect to the angular distribution. The nematic partition function reads

$$Z = \int d(\cos\theta)\exp\left[-\frac{m}{2}\left(3\cos^2\theta - 1\right)\right] \tag{4}$$

m is a mean field parameter *that* represents the strength of nematic interaction. A minimization procedure leads to a relation ship between m, the liquid crystal volume fraction ϕ_l and the so-called Maier-Saupe parameter ν directly related to the transition temperature T_{NI} as $\nu = 4.54\frac{T_{NI}}{T}$. The total free energy is $F = n_T f$ where n_T is the total number of sites. For cured systems involving linear polymeric chains, these equations remain valid but N_2 must be taken proportional to the molar mass.

The formalism is now complete for constructing the phase diagram of uncured monomer mixtures and for systems involving linear polymers. [40, 41] Composition of coexisting phases in a given miscibility gap are determined by equating the chemical potential of each constituent in these phases. Chemical potentials in the isotropic state are simply

$$\frac{\mu_{1i}}{k_B T} = \ln\phi_1 + \left(1 - \frac{N_1}{N_2}\right)\phi_2 + \chi N_1 \phi_2^2 \tag{5}$$

$$\frac{\mu_{2i}}{k_B T} = \ln\phi_2 + \left(1 - \frac{N_2}{N_1}\right)\phi_1 + \chi N_2 \phi_1^2 \tag{6}$$

Nematic counterparts are

$$\frac{\mu_{1n}}{k_B T} = -\ln Z + \frac{1}{2}\nu\phi_1^2 S^2 \tag{7}$$

$$\frac{\mu_{2n}}{k_B T} = \frac{1}{2}\frac{N_2}{N_1}\nu\phi_1^2 S^2 \tag{8}$$

These equations allow derivation of bimodal curves and spinodal curves are given by equating the second derivative of the total free energy to zero. A simple manipulation yields

$$T = 2B\left[\frac{1}{N_1\phi_1} + \frac{1}{N_2\phi_2} - 2A - \frac{\nu S}{N_1}\left(S + \phi_1\frac{\partial S}{\partial \phi_1}\right)\right]^{-1} \tag{9}$$

Results for the uncured monomer NOA65/E7 mixture are shown in figure 4a. We see essentially 2 regions, a wide region where the system exhibits the properties of a homogeneous single phase solution with no effect of anisotropy of the solvent. A miscibility gap with a reduced area consists of large nearly pure liquid crystal droplets exhibiting nematic order are coexisting with a monomer rich phase. This phase diagram was established by optical microscopy and DSC as described in reference. [36] The solid line was calculated according to the model described above. The initial system has the properties of a solution characterized by a viscosity that is sensitive to temperature, composition and anisotropy of the liquid crystal. It is of outmost importance to select properly the conditions of curing in order to achieve the desired characteristics of final products.

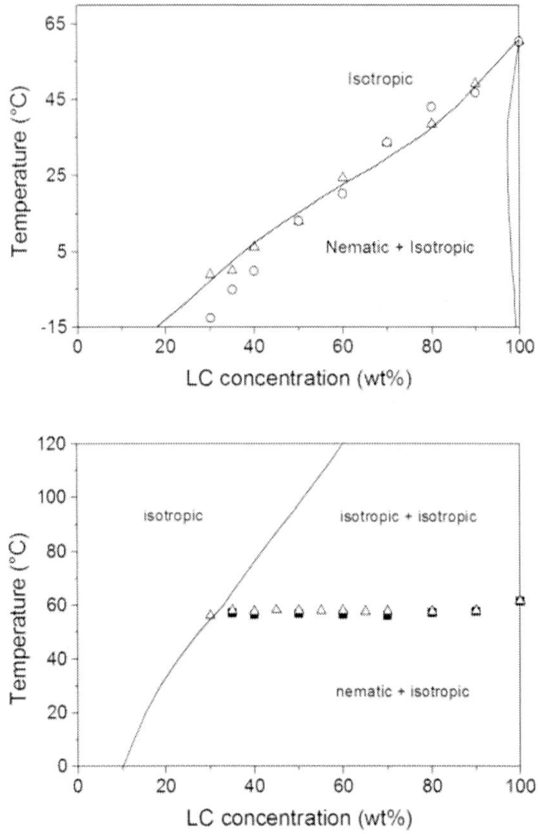

Figure 4. (a) Phase diagram in the temperature versus LC concentration frame for the uncured NOA65/E7 system. Symbols are experimental data (Δ DSC, ○ Polarized Optical Microscopy) and the solid line is the calculated diagram using the models of section 2 with the parameters N1=, N2=,χ=, TNI=60°C. (b) The same as panel (a) for the cured system. The parameter for the calculations are N1=1, Nc=4,α=,β=,ϕ0=, TNI=60°C, χ=.

Cured NOA/E7 networks are highly cross linked and the isotropic free energy contains an additional term due to the elasticity. The presence of cross links opposes strand expansion which was induced by the uptake of liquid crystal molecules under good solvent conditions. According to the Flory-Rehner rubber elasticity theory, the isotropic free energy reads

$$\frac{f(network)}{k_B T} = \frac{3\alpha}{2N_C}\left(\phi_0^{2/3}\phi_2^{2/3} - \phi_2\right) + \frac{\beta}{N_C}\phi_2 \ln\frac{\phi_2}{\phi_0} + \frac{\phi_1 \ln\phi_1}{N_1} + \chi\phi_1\phi_2 \qquad (10)$$

N_C is the number of monomer units between consecutive cross links. The constants α and β are model dependent. The former one is often let to be 1 while β takes different values. In the phantom network model of James and Guth [42], β = 0 while Hermans and Guth [43] suggested β = 1 and Flory's [44] affine network model yielded β = 2/*f*, *f* being the mean functionality of cross linking monomers. The quantity ϕ_0 corresponds to monomer volume fraction prior to curing. Other authors [45] have proposed to make α and β functions of the polymer concentration using $\alpha = 1 - \frac{2}{f}(1-\phi_2)$ and $\beta = \frac{2\phi_2}{f}$. The latter result is consistent with that of Flory in the limit of $\phi_2 = 1$ and James and Guth for $\phi_2 = 0$. The spinodal in the cross linked polymer case becomes

$$T = 2B\left[\frac{3\beta - \alpha\left(\phi_0/\phi_2\right)^{2/3}}{3N_C\phi_2} + \frac{1}{N_1\phi_1} - 2A - \frac{\nu S}{N_1}\left(S + \phi_1\frac{\partial S}{\partial \phi_1}\right)\right]^{-1} \tag{11}$$

The phase diagram calculated using this formalism is given in figure 4b. It is worth while to recall that these considerations are appropriate to PDLCs and should be reformulated in the case of systems exhibiting arrays of polymer and liquid crystal domains (HPDLCs).

Fit of the experimental data sometimes requires that the Flory-Huggins interaction parameter be function of both temperature and composition. The other parameters depend on the rubber elasticity model as mentioned above. The curing process brings about drastic changes in the phase diagram with much wider miscibility gaps above and below 60°C. Data of cured systems need to be completed by observing the transition from isotropic single phase on the left hand side to the gap on the right hand side. This transition is quite difficult to catch especially above T_{NI} since the contrast between the 2 isotropic domains is weak.

3.2. Thermo-Physical

Thermo-physical properties are important for the variety of applications involving nano and micron sized polymer materials. Differential Scanning Calorimetry is the standard tool for characterizing these properties providing transition temperatures and corresponding heat exchanges. Essential properties such as the limit of solubility and the amount of liquid crystal dissolved in the polymer can be deduced from these quantities.

Thermograms in figure 5 show distinct peaks of glass transition and NI transition. Positions of the peaks indicate the temperatures while areas yield the amounts of heat exchanges. Results at different compositions lead to the miscibility limit and the amount of segregated liquid crystal. These results are shown in figures 6a and b in terms of the liquid crystal concentration.

The amount of liquid crystal segregated was deduced from

$$\alpha = \frac{m_{LC}^D}{m_{LC}} = \left(1 + \frac{m_P}{m_{LC}}\right) \cdot \frac{\Delta H_{NI}(\phi_{LC})}{\Delta H_{NI}(\phi_{LC} = 1)} \tag{12}$$

where m_{LC}^D and m_{LC} are masses of liquid crystal segregated and in the sample, m_p is the mass of LC dispersed in the polymer. The theoretical curve in figure 6b was calculated using:

$$\alpha = \left(\frac{100}{\phi_{LC}}\right) \cdot \left(\frac{\phi_{LC} - \phi_c}{100 - \phi_c}\right) \tag{13}$$

where ϕ_c is the limit of solubility of liquid crystal in the polymer matrix obtained from the extrapolation of the $\Delta H_{NI}(\phi_{LC})$ curve as shown in figure 6a.

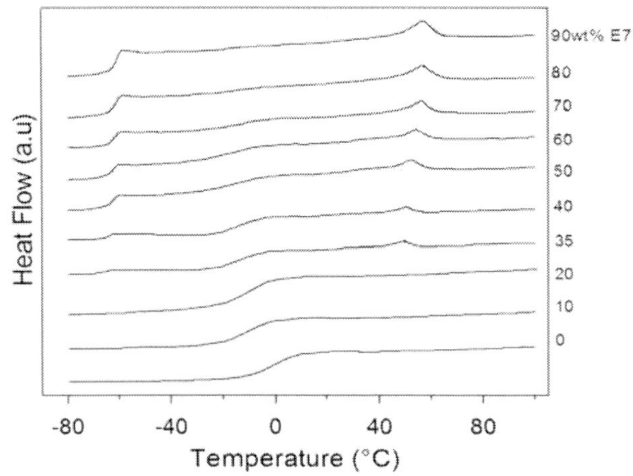

Figure 5. DSC thermograms of the cured NOA65/E7 system covering the whole range of E7 concentration from 0 (pure NOA65) to 90wt% E7 as indicated on the figure.

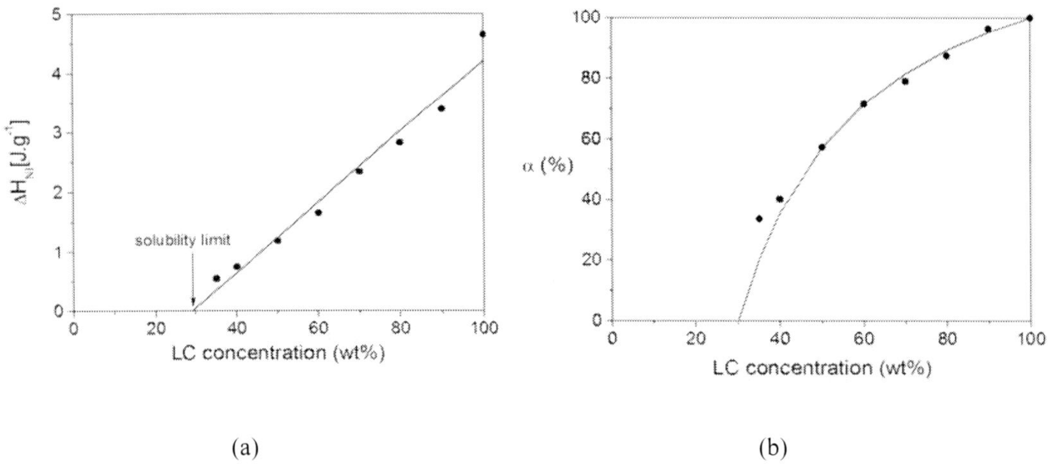

Figure 6. (a) Amount of heat exchange at the nematic to isotropic transition for the cured NOA65/E7 system versus E7 concentration. Note that extrapolation of the line to the x-axis defines the limit of miscibility of E7 in the polymer. The variation of $\Delta H_{NI}(\phi_{E7})$ is linear. (b) Amount of E7 segregated (inclusions) versus concentration for the NOA65/E7 cured system. The variation $\alpha(\phi_{E7})$ is nonlinear.

Thermograms exhibit peaks at the glass transition. The sharp drop of polymer T_g from -20°C (for 10%NOA) to -60°C (for pure E7) is due to plasticizing effect of LC molecules dispersed in the polymer. T_g of E7 remains fixed at -60°C supporting the view that the liquid crystal phase is pure.

4. Thermo-Mechanical and Visco-Elasticity

4.1. Thermo-Mechanical

In this section, we discuss results obtained by thermo-mechanical measurements of NOA65 and liquid crystal systems. The analysis focuses here on the changes of mechanical properties during polymerization with time, temperature and composition. In the static mode, the stress versus strain curves exhibit significant strengthening of the network in the course of polymerization. For pure NOA systems, at short times the variation is linear with a small slope indicating that the mechanical modulus is low. As curing lasts longer and radiation dose gets higher, the curve takes a steeper rising slope and the polymer exhibits higher a mechanical strength with enhanced Young modulus.

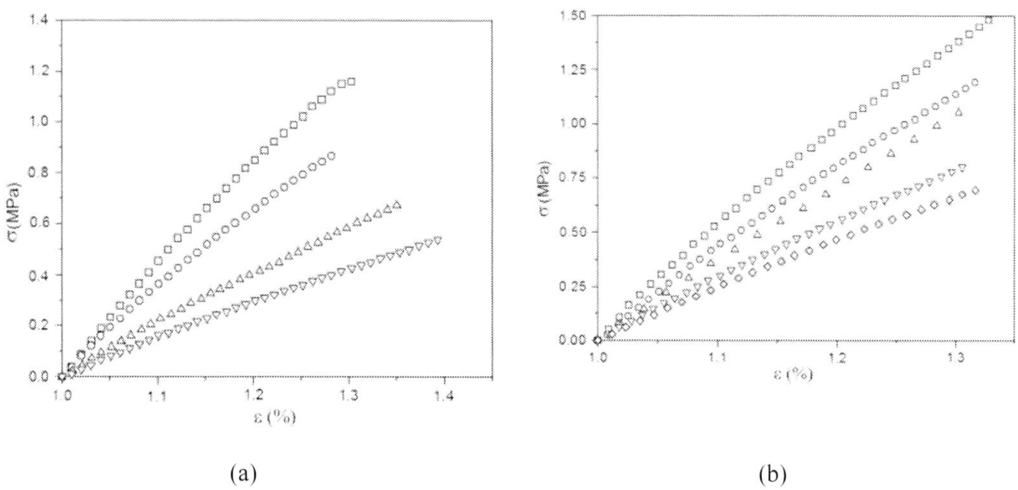

Figure 7. Static mechanical data showing stress-strain curves for NOA65/E7 cured systems (a) with curing time in the ascending order t=5, 10, 20, 30mn, ϕ_{E7}=15wt%. (b) with the content in E7 in the ascending order ϕ_{E7}=35, 30, 25, 20, 0wt% curing time 20mn.

Figure 7a and b illustrate these features at different times of photo-curing and concentrations of the liquid crystal, respectively. Obviously, as the system is exposed longer to the curing radiation, the polymer becomes harder and the increase of stress more significant. Initial slopes of these curves yield the young modulus of the film which is displayed in figure 8.

After nearly 30min of exposure, the modulus reaches a maximum in the presence of 35wt%E7. But this maximum is quite low compared to pure NOA65. In the latter case, the modulus is much higher but takes longer time to reach its upper limit.

The modulus can be used to estimate the molecular weight of a polymer strand between consecutive cross-links M_c. The relationship between M_c and E is simply $M_c = \frac{\rho RT}{E}$, ρ being the polymer density. [46,47] Figure 9 gives M_c in terms of curing time for 3 systems including pure NOA65. In the latter case Mc is lowest and the network is mechanically tougher. As the liquid crystal concentration increases, M_c goes up and the resulting film gets softer. These considerations are also useful for completing the thermodynamic description since M_c is explicitly involed in the elastic free energy according to Flory-Rehner theory of rubber elasticity.

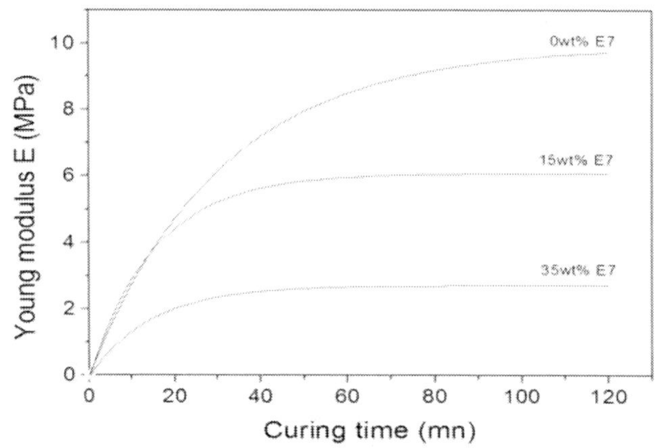

Figure 8. Young modulus E(MPa) versus curing time for NOA65/E7 system at different E7 concentration as indicated on the figure.

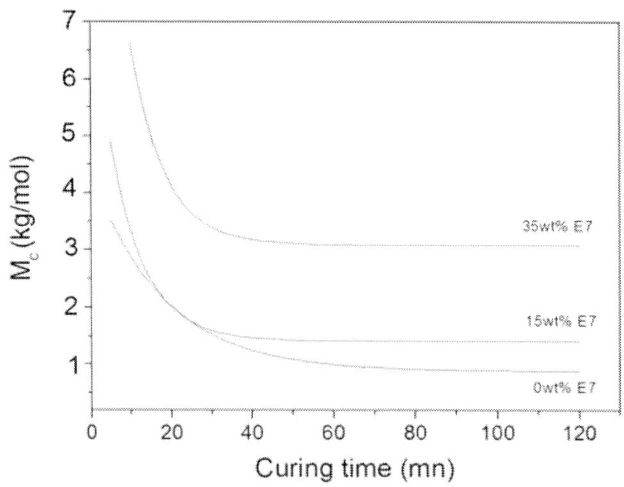

Figure 9. Molecular weight between consecutive cross-links of the NOA network versus curing time at different E7 concentration as indicated on the figure.

The amount of LC that remains in the polymer matrix increases with total ϕ_{LC} and controls a variety of useful properties including mechanical strength, optical transparency, electro-optic responses and visco-elastic properties.

4.2. Dynamical Mechanical Measurement and Visco-Elasticity

When applying a periodic stress with a given frequency, the response functions have an on-phase component related to the elastic modulus and an out-of-phase component related to the loss modulus. The complex representation $E^*(w) = E'(w) + iE''(w)$ $(i^2 \equiv -1)$ characterizes thermal history and dynamical relaxations at the molecular level.

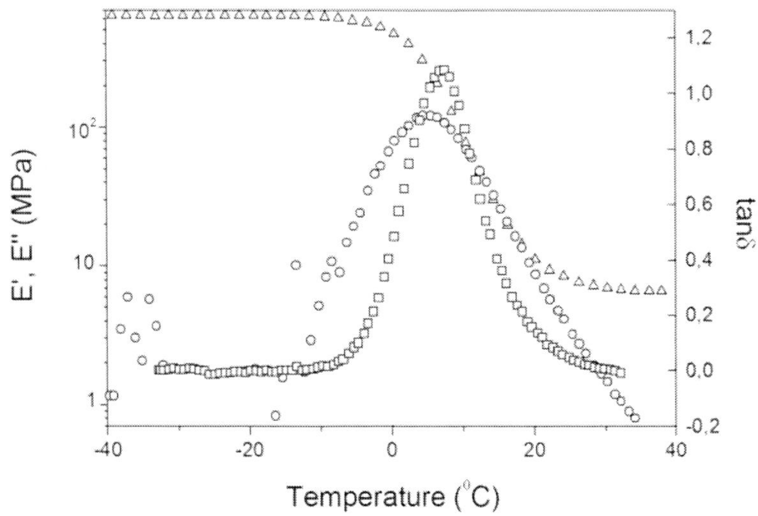

Figure 10. Dynamical mechanical data of pure NOA65 and system showing the elastic E'(Δ) and loss E''(○) modulus and tangent of the loss angle δ (□) versus temperature.

Typical thermo- mechanical data for E', E'' and $tan\delta = E''/E'$ are shown in figure 10. Below T_g, the glassy state is characterized by the same behavior for E', E'' and $tan\delta = E''/E'$. All curves merge together. Above T_g, the rubbery state shows a plateau in the curve of E' at higher levels for densely cross-linked networks. This is expected since the mechanical modulus gets higher. The glass transition temperatures predicted from either E' (drop to the rubbery plateau) E'' (maximum) and tanδ (maximum) show some discrepancies with DSC data. The latter technique often underestimates T_g. It is admitted that thermo-mechanical data are more reliable for locating T_g of highly cross-linked networks but it should be kept in mind that DSC in the dedicated technique for characterizing thermal events such as the glass transition with a reasonable accuracy. Mechanical properties exhibit a sharp transition at T_g which is commonly used to estimate this temperature and complement the measurements made with other techniques such as DSC. But T_g depends on curing time (i.e. radiation dose of photo-curing). For shorter times, T_g is lower and the network softer. As curing process lasts longer, higher cross-linking densities yield tougher networks and the glass transition shifts to higher temperatures. In figures 11 we show the variation of T_g versus curing time for pure NOA65 obtained from DSC and from thermo- mechanical data.

In the dynamical regime, moduli can be expressed in terms of frequency ω and characteristic time constants of normal modes. [46] These moduli obey different scaling laws in terms of ω. For frequencies higher than τ_1^{-1} the inverse time constant of the fundamental mode, the following scaling laws are predicted $G'(\omega) \sim \omega^2$ and $G''(\omega) \sim \omega$. For frequencies lower than τ_1^{-1}, scaling laws are quite different depending on the effects of long range hydrodynamic interactions and the quality of solvent. In the Rouse limit where hydrodynamic interaction is completely screened due to high viscosity of the fluid and one has $G' \sim G'' \sim \omega^{1/2}$ while the effects of long range viscous modes yield $G' \sim G'' \sim \omega^{2/3}$. But such scaling is valid only under good solvent conditions. If excluded volume interaction comes into play and the solvent (here solvent is the LC) has good solvatation qualities, then the scaling behavior becomes $G' \sim G'' \sim \omega^{5/g}$. Precise measurements of visco-elastic properties in a wide frequency range should allow us to distinguish between these scaling laws and conclude about the validity of the model and quality of solvent.

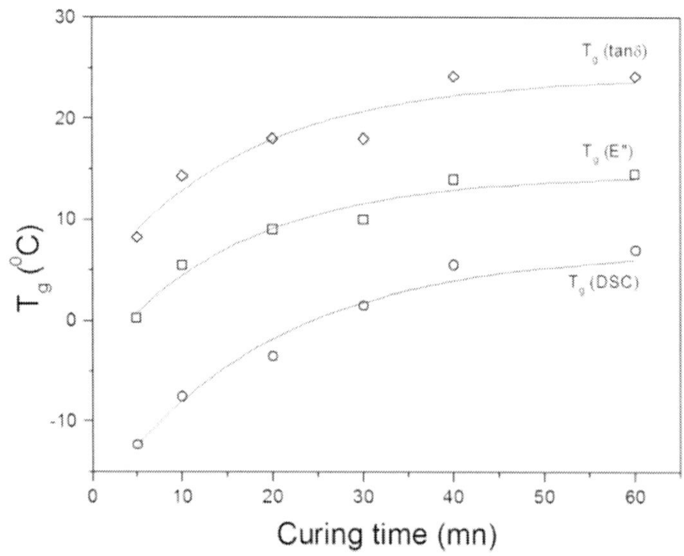

Figure 11. The glass transition temperature T_g for NOA65 in terms of curing time as predicted by DSC and mechanical data. Calorimetric method underestimates T_g while tan yields the highest values.

5. Electro-Optical Responses

When an electromagnetic wave impinges onto an isotropic medium, it propagates with the same speed in all directions function of the index of refraction. When the same wave enters an anisotropic medium such as the liquid crystal E7, it sees 2 different indices of refraction at right angles and splits into two parts. The first part consists of an ordinary wave propagating along the direction of the principal axis of the anisotropic molecules with a speed that is proportional to the ordinary index of refraction. [48] The second part is an extraordinary wave along the perpendicular direction. These two waves undergo a phase lag which is directly related to the optical anisotropy. Measuring changes in the phase lag with the applied signal gives access to changes in the anisotropy and vice-versa. In the case of NOA and E7 composites, there are differences in the indices of refraction at the interfaces in addition to fluctuations due to randomness in the orientation of the liquid crystal directors in the pores. These structural heterogeneities give rise either to diffraction of light or scattering. It is this aspect that is used to probe electro-optic properties of composite materials and promote their applications in sensing devices. From the above analysis, it is clear that morphology and thermo- physical properties should be correlated with electro-optical responses of the systems. In general, these materials consist of films with variable transparencies depending on the conditions of preparation. If pores have micrometer size (i.e. in the range to the wavelength of visible light) one would expect a substantial amount of scattering in the quiescent state due to strong fluctuations at the polymer and liquid crystal interfaces.

These fluctuations are due to abrupt changes in the indices of refraction throughout the medium and to randomness in the orientation behavior of directors distributed along the film thickness. Therefore, the film would show a trans-lucid or opaque texture in the quiescent state.

When applying an electric field, nematic directors orient along its direction and scattering is lowered down significantly leading to transparent texture. This electro-optic sensitivity is used in commutable windows to maintain privacy in selected areas. If the pore size is in the nano meter range, the scattering of light disappears.

5.1. Controlled Light Transmission

The principle of functioning of PDLCs is based upon light transmission controlled via electric voltage with prescribed amplitude and frequency. [49] Under certain circumstances, light is scattered in the off-state when no electric voltage is applied, hence transmittance is low.

If an electric field is applied, and dielectric anisotropy positive the nematic directors are aligned along the field and light transmission is high. To enhance transmission, it would be suitable that the polymer index of refractive be close to the ordinary component of the LC. This condition is nicely fulfilled for NOA65 and the eutectic mixture E7 provided that the amount of LC remaining in the polymer network is low.

The curve of transmission versus voltage has a typical shape shown in figure 12. This variation exhibits a hysteresis phenomenon. A preliminary scan up of voltage indicates that transmission increases with voltage before reaching a maximum. Voltages for which transmission is 10% the maximum value is named V_{10} while that for which it is 90% is called V_{90}.

These data have a practical importance for characterizing the electro-optic performances of the film. Another decisive parameter is the contrast ratio T_{ON}/T_{OFF} where T_{ON} and T_{OFF} denote transmission levels in the ON (electrically activated) and OFF (quiescent) states. In the scan down of voltage, the curve is not superimposed with the original one and hysteresis appears. An evaluation of the hysteresis width is given by ΔV_{50}, the voltage width at half maximum of transmission.

The aimed performances of the composite film are assessed according to the maximum contrast and lowest voltage possible to reduce cost and risks to the material. [48] Additional criteria for assessing the performances include time responses.

Clearly the possibility to control reversibly transmission level with short time responses is desirable. Typically a field of 1V/μm is required to make a film transparent within hundred *ms.* [50]

Response times depend on the nature of liquid cystal and on visco- elastic properties of the whole medium. Anchoring properties at the Polymer/LC interfaces play a determinant role in this context. A rough estimate of the field required to align all liquid crystal molecules along the director of a given droplet depend on the material properties as $E \sim R^{-1}\left(\frac{K}{\Delta\varepsilon}\right)^{\frac{1}{2}}$ where R is mean radius of inclusion, K the Frank elastic constant and $\Delta\varepsilon = \varepsilon_{//} - \varepsilon_{\perp}$ the dielectric anisotropy. This means that the applied electric field is inversely proportional to the liquid crystal domain size and increases with the square root of the ratio $K/\Delta\varepsilon$. Of course, these features should be revisited if the shape of inclusion deviates from spherical symmetry and if one needs to include the effects of electrical resistivities of polymer and liquid crystal. Slight changes in the chemistry, molecular architecture and dimensions of the constituents may lead to different properties with new effect paving the way for other applications.

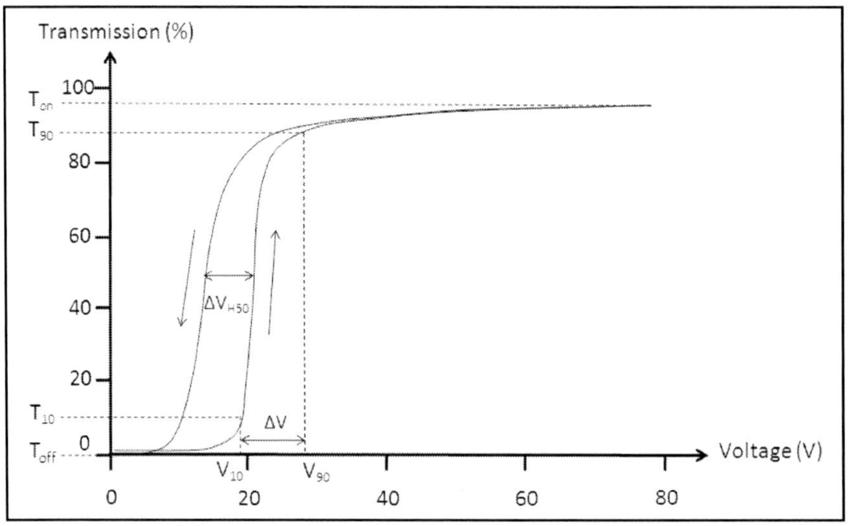

Figure 12. A typical electro-optic curve showing the visible light transmission versus voltage applied for a hypothetic PDLC cell. This figure illustrates the hysteresis form and same relevant electro-optic parameters.

Film thickness and conditions of preparation bear strong impacts on the composite yield in specific application. It is known that transmission π decays exponentially with film thickness (δ), inclusion scattering cross section (σ) and their number N. More specifically, one may write $\ln \pi = -N\sigma\delta$. Increasing the number of inclusions and film thickness leads to multiple scattering. Likewise, films with densely cross-linked networks and smaller liquid crystal domains necessitate higher activation fields and longer response times. All these features affect negatively the electro optical performances with material poor response characteristics.

5.2. Spectral Selectivity

As mentioned earlier, a light beam impinging onto an anisotropic medium splits into 2 waves traveling at different speeds and developing a phase lag proportional to the birefringence Δn.

Applying an electric field, the phase lag is modified providing the possibility to control the spectral selectivity and multiplexing phenomena. These features are illustrated in figures 13 and 14.

The former is obtained from a close examination of the transmission versus wavelength plot exhibiting an oscillatory behavior with relatively weak amplitude.

The oscillations are displaced when voltage is applied enhancing the phase shift between the ordinary and extraordinary waves. This phase shift is directly related to the birefringence Δn which can be controlled by changing the amplitude and frequency of the field. This property is used in designing nano-structured systems for multiplexing and routing of telecommunication signals. A novel method is used to evaluate changes in the birefringence Δn from the spectral transmission consisting of expressing the quantity Λ from figure 13a. [35,51,52]

$$\Lambda = \frac{\lambda_{i+p}^{(1)} - \lambda_{i+p}^{(2)}}{\lambda_{i+p}^{(1)} - \lambda_{i}^{(1)}} \qquad (14)$$

where the wavelengths λ's are defined in figure 13a. Assuming that k_i, the ratio of the optical length and wavelength λ_i is conserved after application of the voltage means

$$k_i = \frac{2n_1 d}{\lambda_i^{(1)}} = \frac{2n_2 d}{\lambda_i^{(2)}} \qquad (15)$$

and

$$k_{i+p} = \frac{2n_1 d}{\lambda_{i+p}^{(1)}} = \frac{2n_2 d}{\lambda_{i+p}^{(2)}} \qquad (16)$$

d is the width of the Perot-Fabry cell (see figure 13b). Substituting those quantities into Eq14 yields

$$\Lambda = 2d\Delta n.\frac{k_i}{2dn_1}.\frac{1}{\left(k_i - k_{i+p}\right)}, \qquad p = k_i - k_{i+p} \qquad (17)$$

Birefringence Δn is calculated according to:

$$\Delta n = \frac{\lambda_i^{(1)} p}{2d} . \frac{\lambda_{i+p}^{(1)} - \lambda_{i+p}^{(2)}}{\lambda_{i+p}^{(1)} - \lambda_i^{(1)}} \quad (18)$$

Figure 14 gives Δn in terms of the voltage applied. The lowest curve corresponds to the first scan up of the voltage. The other two curves describe the standard hysteriesis phenomena observed in electro-optical properties.

This means that voltage scans up and down do not lead to the same values and information is stored.

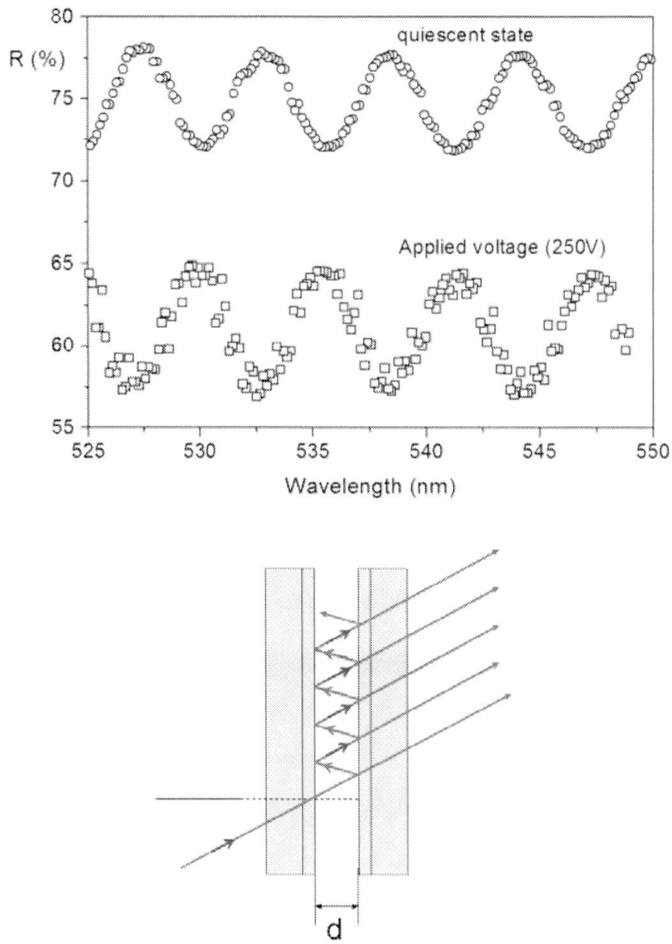

Figure 13. (a) Reflection versus wavelength of NOA65/E7 system under quiescent state (upper curve) subject to an applied voltage of 250V. (b) Perot-Fabry cell.

Orientation distributions of liquid crystal molecules differ slightly and some molecules remain oriented because of interfacial effects at the borders of polymer and liquid crystal domains. Anchoring forces depend strongly on the elastic constants while relaxations of the liquid crystal molecules undergoing electric torques depend on viscosities. Sometimes, memory effects are present when the non activated state is finite.

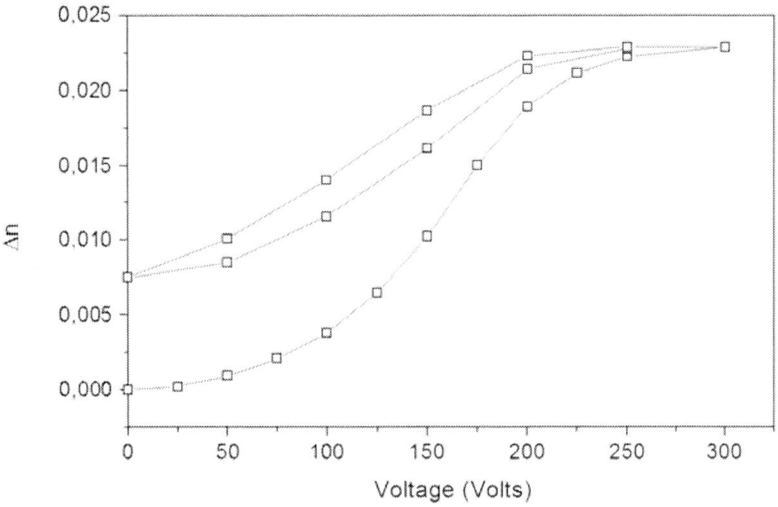

Figure 14. Change in the birefringence Δn versus voltage applied of NOA65/E7 system.

Conclusion

This paper deals with polymer and low molecular weight liquid crystals based on NOA65 and E7. Richness of these composites is highlighted from the fundamental aspects and the broad spectrum of applications at the front edge of modern technologies. Both NOA65 and E7 exhibit complementary properties that make them suitable partners paving the way for many novel developments. The variety of physical properties presented in different sections of this chapter brings a cascade of results that are tightly correlated although involving different processes. Choice of the initial composition and conditions of preparation, control of the kinetics during material formation provide the ability to orient the synthesized product to one or another application. The first section shows how this can be implemented via material synthesis using PIPS. Further development in the theoretical aspect to determine exactly the impact of relevant parameters is still needed to improve the control of interplay between kinetics of polymerization and phase separation. This should help us to enhance the quality of morphology and its development during the course of material preparation at different stages from the initial solution to the final product. A precise knowledge of the phase behavior and thermo- physical properties is crucial since this information explains to a large extend the quality of performances under practical conditions of application. It also enables one to identify the parameters to be modified to enhance these performances. Mechanical and visco-elastic response functions are highly useful for similar reasons. Understanding how the mechanical strength varies and how a dynamic process evolves with time or frequency is crucial from practical and academic points of view. Properties such as T_g can be determined independently from this study and compared with the results obtained using calorimetric techniques. These cross checks maybe recommended to validating the results and enhance the impact of the correlative studies encompassing a variety of typical aspects. Clearly, the applications of our composite materials are based on their electro-optic responses. This investigation shows some research developments to make these materials useful for modern technologies. PDLCs with micron and nano meter sized inclusions offer options for different kinds of applications. Because of their high scattering ability, PDLCs with micron sized inclusions are used mostly in privacy windows with controllable light transmission and display technologies. On the other hand similar systems with nano sized pores are more suitable for telecommunication routing and multiplexing devices. The methodology for investigating the properties of these systems is nevertheless the same.

Acknowledgments

Part of this work was accomplished while the first author was a member of the international max-planck research school (mainz, germany). We are thankful to friends and collegues from the universities of lille and dunkerque (france) and from the mpip in mainz (germany) for many years of fruitful and enjoyable collaboration.

References

[1] Smith, G.W.; Physical Review Letters, 70(2), 198-201 (1993).
[2] Nephew, J.B.; Nihei, T.C.; Carter, S.A.; *Physical Review Letters*, 80(10), 3276-3279 (1998).
[3] Kyu, T.; Nwabunma, D.; *Macromolecules*, 34, 9172 (2001).
[4] Smith, G.W.; *Mol. Cryst. Liq Cryst.*, 196 98 (1991).
[5] Fujikake, H.; Takizawa, K.; Kikuchi, H.; Fyuji, T.; Kawakita, M.; *Jpn J. Appl. Phys.*, 37, 895, (1998).
[6] Sansone, M.J.; Khanarian, G.; Leslie, T.M.; Stiller, M.; Allman, J.; Elizondo, P.; *J. Appl. Phys.*, 67, 4253, (2001).
[7] Lovinger, A.J.; Amundson, C.R.; Davis, D.D.; *Chem. Mater*, 6, 1726-1736 (1994).
[8] Doane, J.W.; in Liquid Crystals: Applications and uses, B. Bahadur Ed. World Scientific River Edge NJ (1992).
[9] Doane, J.W.; Polymer Dispersed Liquid Crystals Displays, World Scientific, Singapore (1990).
[10] Bhargava, R.; Wang, S.Q.; Koenig, J.L.; *Macromolecules*, 32, 8989-8995 (1999).
[11] Nwabunma, D.; Kyu, T.; *Macromolecules*, 32, 664-674 (1999).
[12] Okada, M.; Sakaguchi, T.; *Macromolecules*, 32, 4154 (1999).
[13] de Gennes, P. G.; Liquid Crystals, Clarendon Press, Oxford (1974).
[14] de Gennes, P. G.: Scaling concepts inn polymer physics, Cornell University Press, Ithaca (1979).
[15] Ong, H.L.; Optical Properties of Polymer Dispersed Liquid Crystals, Nonlinear Optical Properties of Liquid Crystals and Polymer Dispersed Liquid Crystals, World Scientific, Singapore (1997).
[16] Crawford, G.P.; Zumer, S.; Liquid Crystals in Complex Geometries, Taylors and Francis, London (1996).
[17] Luchetta, DE. , L. Luchetti, F. Gobbi, S. Simoni,; *Mol. Cryst. Liq. Cryst.*, 360, 93, (2001).
[18] Di Bella, S.; Lucchetti, L.; Simoni, F.; *Mol. Cryst. Liq. Cryst.*, 336, 247-256 (1999).

[19] Nwabunma, D.; Kim, K.J.; Lin, T.; Chien, L.C.; Kyu, T.; *Macromolecules,* 31, 6806 (1998).
[20] Kim, J.Y.; Cho, C.U.; Muhoray, P.P.; Mustafa, M.; Kyu, T.; *Phys Rev Lett*, 71, 2232 (1993).
[21] J Oh, AD Rey, *Macromol. Theory Simul.,* 9, 641 (2000).
[22] Luo, K.; *Eur Phys J.,* 42, 1400 (2006).
[23] Chan, P.K.; Rey, A.D.; *Macromolecules*, 20, 2135 (1997).
[24] Belfoort, B.; Benfoort, D.; Coqueret, X.; *Macromol. Theory Simul.*, 9, 725 (2000).
[25] Mihelic, I.; Kolooini, T.; Podgornick, A.; *J. Appl Polym Sci.,* 87, 2326 (2003).
[26] Qian, T.; Kim, J.H.; Kumar, S.; Taylor, P.L.; *Phys Rev E,* 61, 4007 (2000).
[27] Maschke, U.; Coqueret, X.; Benmouna, M.; *Macromol. Rapid Commun,* 23, 150 (1999).
[28] Date, M.; Takeuchi, Y.; Kato, K.; *J. Phys. D, Appl. Phys.*, 32, 3164 (1999).
[29] Bunning T.J.; Natarjan L.V.; Tondiglia V.P.; Sutherland R.L.; *Annu. Rev. Mater. Sci.*, 30, 83, (2000).
[30] Higgins, D.A.; *Advanced Materials*, 12(4), 251-264 (2000).
[31] Dierking, I.; *Advanced Materials*, 12(3), 167-180 (2000).
[32] Benmouna, R.; Benyoucef, B.; *Journal of Electron Devices*, 5, 142, (2007).
[33] Benmouna, R.; Rachet, V.; le Barny, P.; Feneyrou, P.; Maschke, U.; Coqueret, X.; *J. Polym. Eng.*, 26(5), 499, (2006).
[34] Benmouna, R.; Rachet, V.; le Barny, P.; Feneyrou, P.; Maschke, U.; Coqueret, X.; *J. Polym. Eng.*, 26(7), 655, (2006).
[35] Benmouna, R.; Benyoucef, B.; *Comples*, 36B, 14, (2007).
[36] Benmouna, R.; Benmouna, M.; *Journal of Chemical and Engineering Data*, (2010).
[37] Smith, G.W.; *Physical Review Letters*, 70(2), 198, (1993).
[38] Flory, P.J.; Introduction to Polymer Chemistry, Cornell University Press, Ithaca (1956).
[39] Maier, M.; Saupe, A.; *Naturforsch*, 14a 882 (1959) and 15a 287 (1960).
[40] Benmouna, F.; Bedjaoui, L.; Machke, U.; Coqueret, X.; Benmouna, M.; *Macromol. Theory Simul.*, 7, 599 (1998).
[41] Benmouna, F.; Coqueret, X.; Machke, U.; Benmouna, M.; *Macromolecules*, 31, 4879 (1998).
[42] James, H.; Guth, E.J.; *J. Chem. Phys.*, 15, 669 (1947).
[43] Hermans, J.J.; *J. Polym. Sci.,* 59, 107 (1962).
[44] Grassley, W.W.; *Adv. Polym. Sci.*, 16, 1 (1974).
[45] Petrovic, Z.S.; Macknight, W.J.; Koningsveld, R.; Dusek, K.; *Macromolecules,* 20, 1088 (1987).
[46] Benmouna, R.; Benyoucef, B.; J. *App. Polym. Sci.,* 108, 4072-4079, (2008).
[47] Ferry, J.D.; Viscoelastic Properties of Polymers, John Wiley and Sons Inc., NY (1980).
[48] Binov, L.M.; Chigrinov, V.G.; Electro Optic Effects in Liquid Crystal Materials, Springer, NY (1994).
[49] Doi, M.; Edwards, S.F.; The Theory of Polymer Dynamics, Clarendon Press, Oxford (1986).
[50] Drzaic, P.S.; Liquid Crystal Dispersions, World Scientific, Singapore (1995).
[51] Drzaic, P.S.; *Liq. Cryst.*, 3, 1543 (1988).
[52] Raynes, E.P.; *Phil Trans R Soc. Lond A*, 309, 167 (1983).

In: Adhesive Properties in Nanomaterials, Composites ... ISBN: 978-1-61209-268-3
Editors: K. A. Wilkinson and D. A. Ordonez

Chapter 2

ENVIRONMENTAL AND CHEMICAL DEGRADATION OF BONDED POLYMERIC COMPOSITE JOINTS

Valeria La Saponara*[*1], *Richard A. Campbell*[2], *Patrick Sullivan*[3] and *Douglas Dierdorf*[2]

[1]Mechanical and Aerospace Engineering, University of California, Davis, CA, USA

[2]Applied Research Associates, Tyndall AFB, FL, USA

[3]Air Force Research Laboratory, Tyndall AFB, FL, USA

ABSTRACT

Bonded joints made with polymeric matrix composites are commonly adopted in structural applications where weight is a critical design parameter. They are also key elements in the repair and retrofitting of damaged structures, e.g. aircraft composite skin and reinforced concrete bridge columns. Advances in the design and inspection of bonded polymeric composite joints will therefore improve joints durability, and consequently the safety of composite structures, in a wide range of applications (aerospace, civil, ship, transportation and wind power engineering).

The scope of this chapter is to discuss one aspect of joint durability: chemical and mechanical degradation of the individual components of a typical aerospace joint, i.e. structural epoxy-based adhesive and carbon/epoxy. In a recent research project, these materials were separately exposed to an aggressive environment, consisting of full immersion in water or anti-icing additive (also called fuel additive) or jet fuel or hydraulic fluid. There were simplified laboratory testing conditions: no coatings, no mixing of fluids (i.e. jet fuel and anti-icing additive), no prior thermo-mechanical damage. Gravimetric data, hardness tests and microscopy support the presence of chemical degradation in the adhesive. The use of simple Fickian and non-Fickian two-stage sorption Langmuir models for gravimetric data appears successful for the results of some treatments, e.g. sorption of fuel additive by adhesive. This finding could be used for the purpose of multiphysics modeling of thermo-mechanical degradation of bonded joints. Finally, chemical degradation distinctly appears through Differential Scanning

* Corresponding author. E-mail vlasaponara@ucdavis.edu, phone +1-530-754-8938.

Calorimetry (DSC) and thermogravimetric (TGA) tests: significant changes were encountered when the adhesive was treated in anti-icing additive or hydraulic fluid, while other treatments seem to be much less detrimental for the adhesive. Carbon/epoxy, on the other hand, is impacted at a much lesser rate by fuel additive.

1. Introduction

1.1. Overview of the Content of This Chapter

This chapter introduces recent examples of engineering applications of polymeric composite materials (Section 1.2), a brief discussion on bonded joints and repairs (Section 1.3), and a summary of types of damage that may be encountered in service (Section 1.4).

Afterwards, findings from a three-year long research project are presented, on the chemical, mechanical and thermal degradation of materials used in aerospace bonded joints: epoxy-based structural adhesive and carbon/epoxy. These were conditioned in contaminants typical of aerospace operations (jet fuel, fuel additive, hydraulic fluid, water), under simplified laboratory conditions that provide a reference frame.

This study is articulated in four main sections: experimental set-up and results (Sections 2-3), one-dimensional analytical modeling of the diffusion/diffusion-reaction processes (Section 4), thermogravimetric and Differential Scanning Calorimetry scans (Section 5). A multi-faceted view is offered of the different stages of vulnerability of these samples. In particular, it is shown that the epoxy structural adhesive may be very susceptible to exposure of hydraulic fluid or fuel additive. For hydraulic fluid treatment, the loss of hardness parallels the drastic volatilization and mass loss of the adhesive at temperatures as low as 100-200 deg. C, when compared to more than 350 deg. C for the control material. This suggests possible damage at normal operating conditions and dramatic consequences in case of a composite fire incident.Further work by our research groups will discuss flammability characteristics of selected adhesive and carbon/epoxy specimens subject to chemical and environmental degradation.

1.2. Applications of Composite Materials

Fiber-reinforced polymer composite structures have been introduced since the 1970s in applications where weight is a major design parameter, starting with aerospace engineering. Airbus pioneered the use of composites in the early 1970s with the A300 series, but the upcoming Boeing 787 will be the first commercial airliner where composites make 50% of the aircraft structural weight. Military fixed- and rotary-wing aircraft have consistently used composites, e.g. in the F-117 Nighthawk, F-22 Raptor, B-2 Spirit, RAH-66 Comanche, V-22 Osprey, and the upcoming F-35 Lightning II and Sikorsky CH-53k. Use of fiber-reinforced polymer composites (also known as fiber-reinforced plastics, FRPs) is also increasing in civil engineering. As of 2003, over 300 composite bridges were reported as present in US alone. [1] FRPs are not only used for new constructions (beams, columns, slabs, walls, pipes, tanks), but also to retrofit existing infrastructure (e.g. for blast mitigation, to replace corrosion-prone steel reinforcement, for seismic retrofit). Two examples of FRP applications in bridges are

Miyun bridge, the first FRP vehicular bridge in the world, which opened in China in 1982, and the retrofit of the I-80 State Street Bridge in Salt Lake City, USA, in the early 2000s. Additional examples are included in recent literature (to name a few, [2-6]).

Moreover, composites are becoming the first choice for the hulls of vessels requiring low weight and low magnetic signatures, e.g. minecountermeasure vessels (e.g. the 72 m Swedish Visby Corvette, launched in 2000, or the new all-carbon/epoxy 27 m US Navy M-80 Stiletto, introduced in 2006).

One of the largest applications of composites worldwide is in wind power, in particular wind turbine blades: for example, in 2007 approximately 44,000 turbine blades were manufactured, corresponding to ~192,000 tons of composite structures. [7] Wind energy could potentially generate 20% of US electricity by 2030. In 2008, there were $US 17 billions invested and 35,000 new jobs opened, in US only. [8]

Moreover, companies such as General Electrics are moving towards the adoption of composite materials for automotive and railway applications, e.g. incorporating thermoplastic semi-structural panels in the passenger cars of Metro-North Railroad, the third largest commuter railroad in North America. [9] Additional worldwide examples are the glass/sandwich constructions for the cars of Las Vegas monorail, the reinforced polyester and phenolic resins for the front-ends of the French TGV and the German InterCity Express trains, and the glass/ and carbon/polyester bodies of the Copenaghen (Denmark) metropolitan trains. [10] Since 2007, South Korea is developing a tilting train with a composite carbody.

1.3. Bonded Joints and Bonded Repairs

Composites may be joined to other composites or to metals through bonding, fastening/riveting, or a combination of those. This holds true also for repairs. Joints, whether bonded or fastened, '*are perhaps the most common source of failure in aircraft structures*'. [11] Bonded joints and repairs may be preferred to fastened joints and repairs for a variety of reasons, such as weight, the need to connect different materials and avoid potential galvanic corrosion, the requirement of smooth surfaces, reduced risk for delamination. However, bonded joints cannot be disassembled, and have stringent requirements for surface preparation, which may not be fully respected. If the bonded joint surface is adequately prepared, the joint will tolerate a certain number of disbonds, and possibly not require repairs. [12]

It has been argued that unnecessary repairs of a well-prepared joint may destroy the existing surface protection (primer, etch, anodize etc.) against chemical attack, and thus leave the restored joint more vulnerable to a hostile service environment. [12] Extensive resources are available on the design, analysis and durability of composite joints and repairs for a wide variety of applications (e.g. [13-23], to name a few).

Examples of bonded joints are in the rudder of the Airbus A300 series, the connection between girders and hulls in composite ships, the connection between reinforced concrete structures (bridges, buildings, chemical plants, marine and water front constructions) and the FRP wraps used to strengthen them, and also between wind turbine blades and the metal propeller hub. Because of the widespread nature of joints in engineering structures, there are strong safety implications stemming from the durability of these joints.

1.4. Challenges

Composites exhibit complex damage modes such as delamination, fiber cracks, matrix cracks, fiber/matrix interface debonding, etc. In service, these modes may be triggered by factors such as thermo-mechanical fatigue, impact damage, environmental and chemical damage, manufacturing defects, overheating, or a combination of all these factors ([5, 15, 24], etc.). The cost of inspection has been estimated to be about 30% of the total cost of the acquisition and operation of composite structures. [25]

In addition, one of their most challenging aspects is composites' behavior when they are on fire, [26-29]: the polymeric matrix material decomposes over a range of temperatures between 350 deg. C to 600 deg. C by releasing heat, soot and a variety of toxic fumes, from carbon monoxide to hydrogen chloride and hydrogen cyanide. While the reinforcement material may be non-flammable (glass and carbon), it will oxidize upon sufficiently high heat, and release inhalable particles. Fire retardant epoxies and phenolic resins are increasingly applied, but they are expensive, and they still release smoke and fumes. Moreover, the polymeric composite may soften and loose load-bearing capability and buckling strength for temperatures larger than 100-200 deg. C. The degradation due to exposure to fire has been indicated as '*the single greatest impediment to the use of FRP composites in the design and construction of advanced ship systems and ship structures*'. [30] Degradation mechanisms that affect the polymeric resin are likely intertwined with the structure's resistance to fire.-

2. Degradation Study

2.1. Experimental

For this work, several specimens were prepared to obtain the diffusion properties through gravimetric tests in 100% relative humidity (100% RH, i.e. full immersion). Mass gains were measured from a) conditioned and control lap joints prepared with structural adhesive and woven carbon/epoxy,b) conditioned and control structural adhesive samples, and c) conditioned and control woven carbon/epoxy samples. There were five specimens per condition, plus the baseline (control). This section introduces materials and test set-up reported in. [31]

2.1.1. Lap Joints Investigation

The very first batch of specimens of this study consisted of single lap joints, the simplest type of bonded joints found in the engineering practice and in the literature. [31]The adherends consisted of T-300 woven carbon/epoxy with weight/area of the dry fabric equal to 98.3 g/m^2.The reinforcement material was infused with degassed Proset® 117 LV epoxy mixed with Proset® 237 hardener. Vacuum Assisted Resin Transfer Molding (VARTM) was used, with a cure cycle of four hours at 50 deg. C followed by sixteen hours at 60 deg. C.

The structural adhesive consisted of a commercial epoxy-based adhesive (Hysol® EA 9360), a two-component toughened paste adhesive with high peel strength, approved for aerospace maintenance, repairs and operations, MRO. [34-35] Epoxies are the most common type of resin for aerospace adhesives and composite applications, e.g. [15], hence it is believed that this adhesive is a sound representative of a material used in the engineering practice (in particular, aerospace operations).

The lap joints had a 25.4 mm overlay area, a 120 mm gauge length, and ten layers of carbon/epoxy overall. [31] The edges of the adherends and adhesive were not sealed. All specimens were tabbed with G10 fiberglass prior to conditioning, to prevent slipping during the tensile teststhat would follow the treatments.Gravimetric tests were followed by static tensile tests. The mass gain was measured by a Mettler balance with a 100 g range and a resolution of 0.1 mg. Five specimens per condition and the control specimens were immersed in open containers of different fluids: fresh water, or jet fuel, or Skydrol® 500B hydraulic fluid, or Prist® Hi-Flash™ anti-icing fuel additive (also called simply 'fuel additive' herein), at room temperature only. Details of these fluids are given in the following section.Due to a time constraint, the immersion tests of the lap joints lasted only 243 hours. This period was not sufficiently long to reach steady-state conditions, but it is similar to testing periods appearing in the specifications of some commercial adhesives. [31] The preliminary results from these lap joints tests showed that the specimens absorbed anti-icing additive at a much higher rate than the other fluids. Also, it was not clearwhat the individual uptakes of adherends, adhesive and tabs were. Therefore, the study continued and focused on the behavior of the individual components of the lap joints, as discussed below.

2.1.2. Adhesive Investigation

Specimens with dimensions 60 mm x 60 mm x 1 mm were prepared and immersed in tightly sealed containers with the same fluids of the preliminary lap joint study: fresh water, or jet fuel, or Skydrol® 500B hydraulic fluid, or Prist® Hi-Flash™ anti-icing fuel additive. The adhesive specimens did not have any commercial coatings, surface treatments, repair patches, which would be more representative of service conditions, but would greatly complicate interpretation of any results (e.g. [32]). The specimens' size followed the ISO 62 standard for sake of manufacturing ease, while the ASTM D5229 standard dictated the rest of the procedure. Conditioning also took place at three different temperatures: room temperature, 70 deg. C and 85 deg. C. The cure cycles of the adhesive and the reasons behind the temperature selection are discussed below, after presenting details on the fluids.

The hydraulic fluid of this study is a phosphate ester-based hydraulic fluid with improved resistance to heat. Hydraulic fluids have been shown to be detrimental to aerospace composites in service, as they attack paint/sealant and resin, [15], and were the likely cause of delamination between honeycomb core and fiberglass composite skin in the rudder surfaces of some Airbus A300 aircraft. [33]

Fuel additive is typically mixed with jet fuel to prevent water freezing in the fuel lines. The additive adopted in this project is diethylene glycol monomethyl ether, and was not combined with jet fuel to simplify the analysis of results. It is possible that the outcome may have been mitigated otherwise. Rider and Yeo, [36], found that diethylene glycol monomethyl ether, which was used in the F-111 of the Royal Australian Air Force, caused '*massive disbonds*' (up to 75% of surface area) in adhesive joints, and softened the adhesive. To accelerate or amplify the adhesive degradation, their tests considered a concentration of diethylene glycol monomethyl ether which was four times greater than normal levels found in field fuel lines. [36]

The current work provides further evidence in support of their worrisome conclusions. The adhesive was cured following the manufacturing guidelines, which recommend curing at room temperature for 5 to 7 days '*to achieve normal performance*'. [35] In addition, it is stated in the specifications that '*Accelerated cures up to 93 deg. C (for small masses only) may be used as an alternative. For example, 1 hour at 82 deg. C will give complete cure*'. A small group of adhesive specimens received post-curing at 82 deg. C for two hours (herein indicated as 'complete cure'), while the specimens cured at room temperature only are described as 'normal cure' specimens. It is believed that the manufacturer recommendation on post-curing may not be followed in situations where hot blankets or full-scale autoclave are not an option, e.g. for very large structures and/or for field repairs.

The stage of partial cure of the 'normally' cured coupons was witnessed by their larger vulnerability to degradation with respect to the post-cured samples, as indicated by their typical lower glass-transition temperatures, T_g (Section 5), and the increase of hardness over time (aging) of the control specimens. [31]

After curing, the conditioning of the adhesive specimens took place not only in the aforementioned fluids, but also with no humidity control and at different temperatures: room temperature, 70 deg. C and 85 deg. C. Two ovens were used: a furnace/oven Series 3710 from Wagner Instruments (for tests at 85 deg. C in 2007), and an internally desparked safety oven, model VWR 1330 (for tests at 70 deg. C in 2008 and 2009).

The two temperatures of 70 deg. C and 85 deg. C were selected based on the service temperature of the adhesive. According to the guidelines of ASTM D5229, gravimetric tests must be run at temperatures no higher than 25 deg. C below the T_g. The adhesive's T_g varies according to the adhesive's state of cure, and its measurement has some dependence on test methodology and temperature rate. It was decided to use the published service temperatures of 107-121 deg. C as a ceiling, in an attempt to study degradation in realistic service conditions. The temperatures of 70 deg. C and 85 deg. C are consistent with aerospace operations: for example, '*Jet fuel is a fluid that will always be in contact with the composite on a long term basis and at elevated temperatures*' (121 deg. C and 177 deg. C in [37]). The temperature of 70 deg. C is '*a typical temperature that an aircraft surface could reach on the ground due to solar heating*'. [32] This research endeavor was not intended as an accelerated aging project to extrapolate long-term behavior and calculate the Time-Temperature Superposition Principle. In fact, different levels of degradation were encountered by the adhesive at 70 deg. C and 85 deg. C, as indicated by varied texture and filler exposure, hardness changes (Section 3), and thermogravimetric results (Section 5). The time duration of the experiments was dictated by the need to reach steady-state moisture uptake or 'apparent' equilibrium (to be discussed in Section 4).

It is possible that the pressure build-up in the containers may have influenced the sorption process, but there is no available data at this point on this effect, in the current tests.

After the tests were stopped, hardness measurements were taken on the dried specimens. Due to the dimensions of the plates for gravimetric work, testing of mechanical properties was challenging, hence the choice of hardness measurements. A first group of specimens was investigated with a Rockwell hardness tester, until a Shore D durometer was purchased, which is more appropriate for plastics. Statistical modeling was carried out for the conversion from a) the Rockwell scale to Shore D scale, and b) the conversion of Shore D to tensile strength, built from another group of materials of known strengths ('calibration materials', [31]). The methods used for these tasks include Shapiro-Wilk normality tests, LOWESS (LOcally WEighted Scatter plot Smoothing) fits, residuals plots and box plots. [31]

2.1.3. Carbon/Epoxy Investigation

Thin specimens were prepared with as-received carbon T-300 plain weave from Sigmatex, infused with Proset® 117 LV epoxy and Proset® 237 hardener. Samples had nominal dimensions 100 mm x 100 mm x 1 mm (ASTM D5229). The measured material properties were equal to $E_{11} = E_{22} = 48.4 \pm 3.71$ GPa, $G_{12} = 2.685 \pm 0.198$ GPa, $\nu_{12} = 0.0380 \pm 0.0153$ for a single lamina. Vacuum Assisted Resin Transfer Molding was used, with a cure cycle of four hours at 50 deg. C followed by sixteen hours at 60 deg. C. Specimens were immersed in the same fluids mentioned above, at room temperature and at 70 deg. C (with the exception of jet fuel conditioning, which took place only at room temperature). Hardness was measured with a Shore D durometer on the dried specimens. The uptake results were not normalized with respect to the weight fraction of the matrix.

3. Mass Gain and Hardness Results

The specimens exhibited different degrees of swelling, color and hardness changes, Figures 1-3. A limited number of microscope pictures highlighted the existence of significant texture variations, swelling and cracking with partial exposure of fillers present in the epoxy adhesive, which were not visible in the control specimens (Figure 2, [31]).

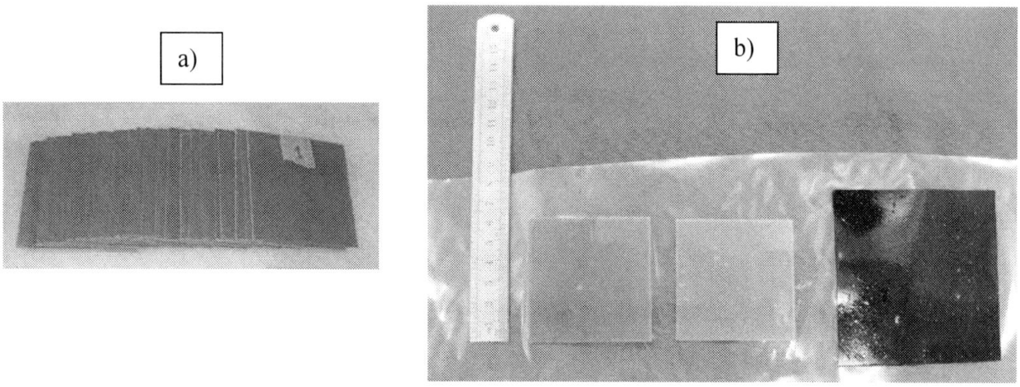

Figure 1. a) Control adhesive specimens with size 60 mm. [31] b) Adhesive specimens conditioned in (left) water at 70 deg. C, (center) hydraulic fluid at room temperature, (right) hydraulic fluid at 70 deg. C.

Mass gain was computed as $m_t(t) = \frac{W_i(t) - W_0}{W_0} \text{x} \, 100$. $W_i(t)$ is the weight at time t of the conditioned specimen, while W_0 is the weight of the same specimen after pre-conditioning and immediately before immersion. Structural adhesive specimens treated in hydraulic fluid gained up to ~175% mass (at 85 deg. C), while those in fuel additive increased their mass of ~85% (at both 70 deg. C and 85 deg. C).

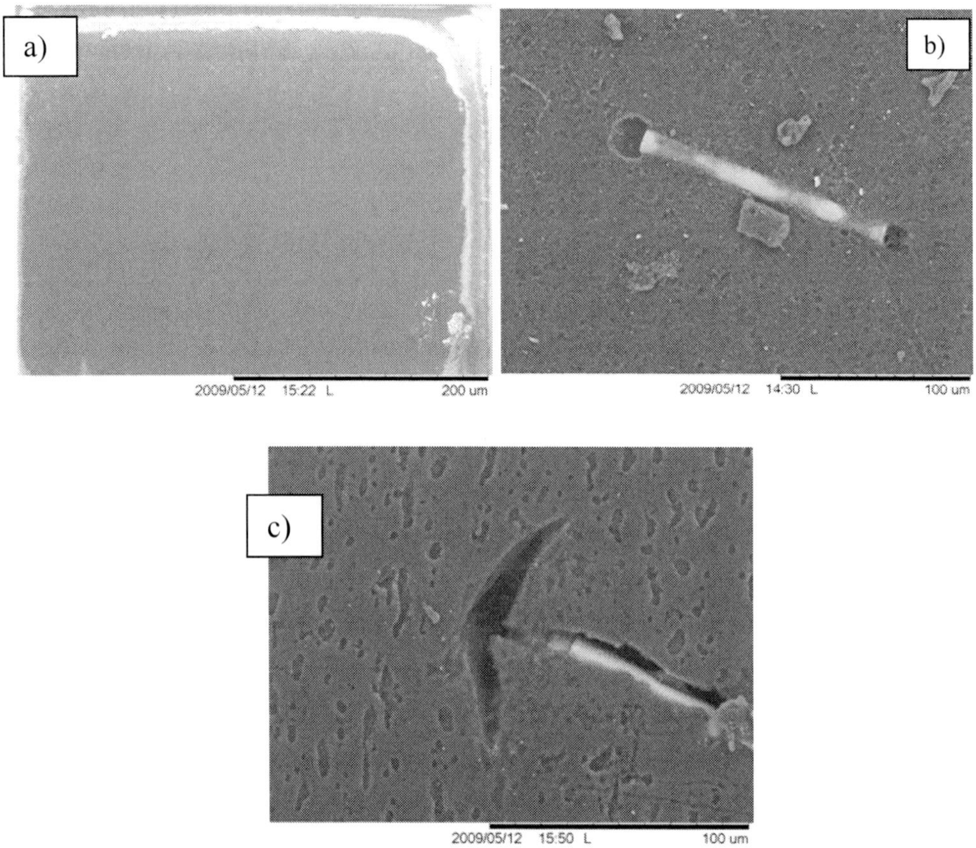

Figure 2. Microscopic images, [31], of: a) Adhesive control specimen at 70 deg. C, b) Adhesive specimen treated in anti-icing additive at 70 deg. C, and c) In hydraulic fluid at 70 deg. C. The rods may be fillers exposed by the chemical reactions. The microscope is a Hitachi TM-1000.

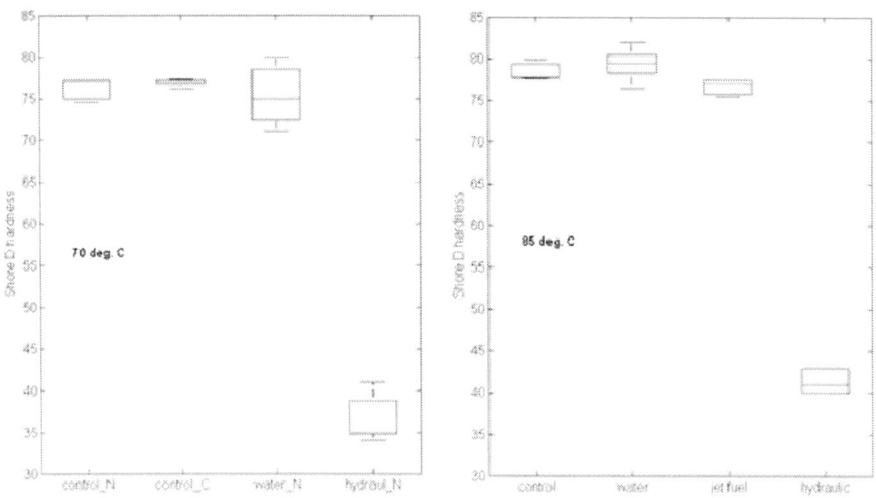

Figure 3. Hardness results for adhesive samples at high temperature (left: 70 deg. C tests; right: 85 deg. C tests). The specimens conditioned in anti-icing additive disintegrated, and their hardness is not available. 'N' and 'C' stand respectively for normal (no post-curing) and complete cure (post-curing).

Note that the median mass gain of the adhesive specimens in water at all conditions was between ~5% and ~7%, which is consistent with the literature, and shows the testing methodology to be appropriate. The carbon/epoxy samples had the largest swelling when immersed in anti-icing additive, 11% (at 70 deg. C). Based on this information and on thermogravimetric results (discussed in Section 5), it is believed that there may be irreversible chemical damage in the adhesive. Section 4 gives details on the exposure time and on simple diffusion models available in the literature, which may be sufficient, in some cases, to capture the overall fluid uptake, even under complex diffusion-reaction conditions.

Hardness dropped considerably for adhesive specimens treated in hydraulic fluid (~50% between median of control and median of conditioned samples, at 70 deg. C and 85 deg. C). Specimens in fuel additive broke into pieces at high temperature, and their hardness could not be reliably measured, Figure 3. The influence of other fluids on the adhesive's hardness did not appear statistically significant. Through LOWESS and least-squares fitting, an equation for the tensile strength was computed in terms of Shore D hardness, [31]:

$$\text{tensile\_strength} = -85.824 + 1.8276 \cdot \text{ShoreD} \text{ , in MPa,} \qquad (1)$$

This equation is an approximation for plastics with strengths up to 80 MPa. From Equation 1, a 50% hardness decrease would correspond to a complete loss of tensile strength.

Among the conditioned carbon/epoxy specimens, the largest change of hardness (~12.5%) is experienced for the treatment in anti-icing at room temperature, Figure 4. A correlation with tensile strength is not available for carbon/epoxy.

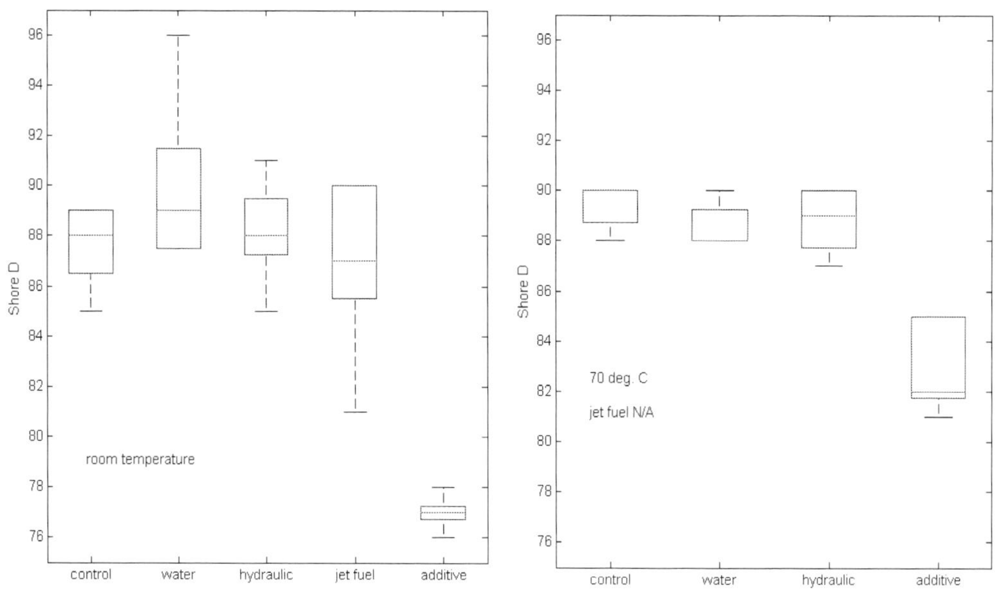

Figure 4. Hardness results for carbon/epoxy samples at room temperature and at 70 deg. C. Tests for jet fuel conditioning are not available at 70 deg. C.

4. Results of Fluid Sorption, Fickian and Langmuir Diffusion Models

4.1. Overview of Fluid Uptake Models

Mass and heat transports cause structural changes in the polymer (swelling, microcavity formation, primary and secondary phase transitions, etc.) that lead eventually to rearrangements of the polymer chains. In the test data, there were cases of moisture uptake which could be modeled by Fickian diffusion (structural changes are dominated by diffusion) or non-Fickian two-stage sorption (diffusion and relaxation have comparable rates, and there is a way to account for both reversible and irreversible sorptions). In these latter situations, apparent equilibrium was observed, i.e. there were several instances of weight increase after intervals at constant weight. In the literature, apparent equilibrium was observed early on by Bagley and Long, [38] for the uptake of acetone by cellulose acetate films, and suggests a combination of diffusion and relaxation phenomena occurring at comparable time scales (e.g. [39-41], etc.). In particular, moisture diffuses in the polymer until it gets saturated and reaches apparent equilibrium. However, structural changes follow slowly, and a new final equilibrium is reached when the structure of the material is saturated, and is also in equilibrium with those boundary conditions. [39]

Fickian diffusion, [42], was a reasonably accurate model for the median weight gain of the samples in the following situations:

a) adhesive samples treated in water (at room temperature for ~1,800 hours, and at 70 deg. C for 620 hours) or in jet fuel (at room temperature for ~1,000 hours), or in anti-icing additive (at room temperature, for ~6,700 hours)
b) carbon/epoxy samples treated in water (at room temperature and at 70 deg. C, for up to ~1,800 hours), or in hydraulic fluid (at room temperature, for ~1,800 hours), or in anti-icing additive (at 70 deg. C, for ~8,600 hours)

A Langmuir model, [43], discussed below, approximated reasonably well the behavior of the median weight gain of:

a) adhesive samples in water (at 85 deg. C) for 620 hours, or in anti-icing additive (at 70 deg. C, for ~13,300 hours, and at 85 deg. C, for ~5,500 hours)
b) carbon/epoxy in anti-icing additive (at 70 deg. C, for ~8,650 hours) or in hydraulic fluid (70 deg. C, for 780 hours)

The mass gain of the adhesive conditioned in hydraulic fluid requires a more complex model, which is work in progress.

5.2. Diffusion Models

Mass uptakes of the median specimen in each group were analyzed with one-dimensional Fickian diffusion and a Langmuir model for two-stage sorption, following the work by Carter and Kibler. [43] Specimen edge effects are neglected.

Fickian diffusivity is calculated from:

$$D_{Fick} = \pi \cdot \left(\frac{2\delta}{4m_{\infty}} \right)^2 \left(\frac{m_{linear}}{\sqrt{t_{linear}}} \right)^2 \quad (2)$$

where m_{linear} and t_{linear} are respectively the percent mass gain and time interval (in seconds, starting at time = 0) of the linear portion of the gravimetric curve, calculated with a conventional least squares fit. Moreover, a widely used approximation was adopted, [44]:

$$m_t \cong m_{\infty} \left[1 - \exp\left(-7.3 \left(\frac{D_{Fick} t}{s^2} \right)^{0.75} \right) \right] \quad (3)$$

s is equal to the thickness 2δ, for a material exposed on two sides to the fluid, and m_{∞} is the percent weight change at saturation.

Langmuir model has been cited/utilized by many researchers to date, e.g. [40-41, 45-50] This approach assumes that the diffusivity D_{γ} does not depend on concentration or on stress, and is due to the interaction of n mobile molecules per unit volume (a way to model a reversible interaction), and N bound molecules per unit volume (irreversible interaction), each with their probability per unit time (respectively, γ and β). At equilibrium, $\gamma \cdot n_{\infty} = \beta \cdot N_{\infty}$, which follows Langmuir's adsorption theory. The physical meaning and actual occurrence of bound and mobile molecules in the sorption process have been discussed by some authors. [45, 48] Langmuir moisture uptake is calculated as:

$$m_t \cong m_{\infty} \left\{ \frac{\beta}{\gamma + \beta} e^{-\gamma t} \left[1 - \frac{8}{\pi^2} \sum_{p=1}^{\infty(odd)} \frac{e^{-\kappa p^2 t}}{p^2} \right] + \frac{\beta}{\gamma + \beta} \left(e^{-\beta t} - e^{-\gamma t} \right) + \left(1 - e^{-\beta t} \right) \right\},$$

under condition that $2\gamma, 2\beta << \kappa$ (4)

In this equation, m_{∞} is the percent weight change at equilibrium (at saturation or at apparent equilibrium) of the median specimen among the five specimens tested per condition. κ is calculated from the diffusivity coefficient and the square of the specimen thickness 2δ: $\kappa = \frac{\pi^2 D_{\gamma}}{(2\delta)^2}$. The methodology for the current work involved the use of two numerical derivatives calculated with a MATLAB published routine (a central difference routine, which is applied to data points at unequal steps from each other, [51]). In fact, mass measurements were taken at unequal time intervals. Moreover, three conditions were to be verified, which were related to short-term and long-term exposures and are the foundations of Langmuir model (see [50], for details).

In some cases, these three conditions could not be satisfied for the median specimen selected for this analysis. Hence, Langmuir fitting curve was not computed for those medians. This should not be necessarily construed as inadequacy of the Langmuir model per se, and could depend on the ill-posed behavior of derivatives.

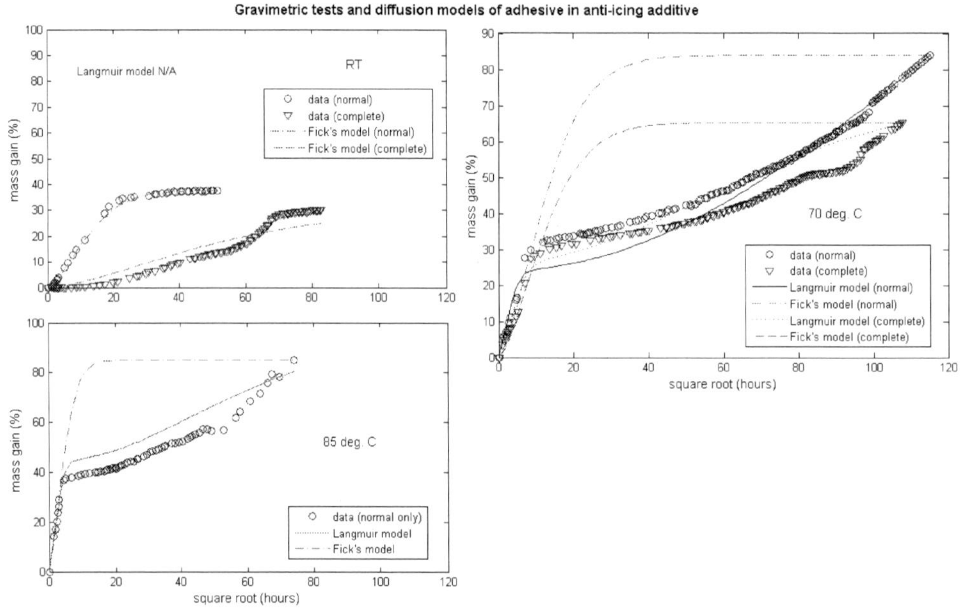

Figure 5. Application of Fickian and Langmuir models to median gravimetric data of structural adhesive in 100% RH anti-icing additive, at room temperature ('RT'), 70 deg. C and 85 deg. C. Oven malfunction caused the interruption of the tests at 85 °C between 2100 and 2600 hours, i.e. 45 and 50 square root(hours).

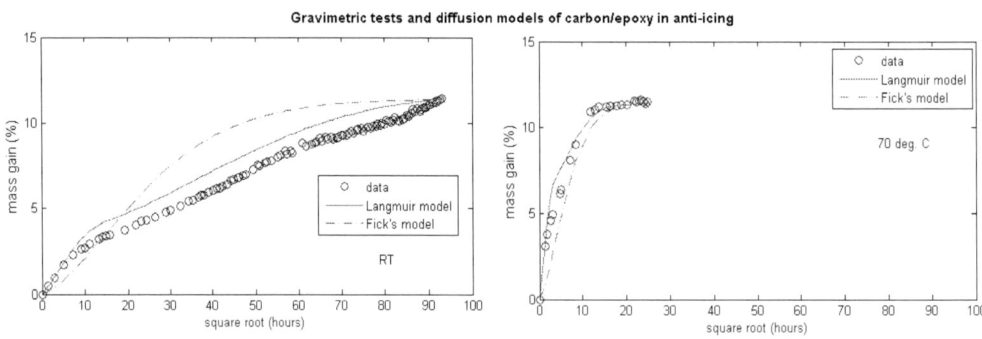

Figure 6. Fickian and Langmuir models to median gravimetric data of carbon/epoxy in anti-icing, at room temperature ('RT') and 70 deg. C.

One such case of unsuccessful Langmuir fit is the adhesive's moisture uptake of hydraulic fluid or jet fuel. However, Langmuir model was reasonably successful in capturing the non-Fickian sorption behavior of adhesive (Figure 5) and carbon/epoxy (Figure 6) in anti-icing additive, which turned out to be the most detrimental fluid in this research project, within the limits and approximations of this project (Section 5).

In conclusion, it was demonstrated that, regardless their limitations (in particular, no dependence on concentration and on temperature), these simple one-dimensional diffusion models were sufficient to well fit experimental data from selected treatments of structural adhesive and carbon/epoxy composites, including immersion in fuel additive. However, further work is needed to investigate the sorption of adhesive in hydraulic fluid. This outcome may help researchers adopting multiphysics commercial or in-house software to predict degradation of structural adhesive and/or bonded joints in service.

5. Results from Differential Scanning Calorimetry and Thermogravimetric Analysis

Differential Scanning Calorimetry (DSC) and Thermogravimetric Analysis (TGA) were performed with a Netzsch Thermogravimetric Analyzer/Differential Scanning Calorimeter (TGA/DSC) Model STA 409 PC, with Al_2O_3 pans and lids. Temperatures ramped for all samples from room temperature (approximately 22 deg. C) to 900 deg. C, at a rate of 15 deg. C per minute. All samples were processed under a nitrogen purge of 50 cc/mi, and were tested towards the end of this three-year project, after appropriate pre-conditioning.

Table 1. DSC data for structural adhesive (rate of 15 deg. C/min).

Treatment/cure type/ test temperature (deg. C)	Transition range (deg. C)
control/normal/room	onset 59, peak 65.5
control/complete/room	onset 87.0, inflection 92.0
control/normal/70	onset 95.2, inflection 101
control/complete /70	onset 93, inflection 98.5
control/normal/85	onset 83.5, peak 92.0
water/normal/room	onset 77.5, peak 86.4*
hydraulic fluid/normal/room	onset 91.1, inflection 94.7
fuel additive/normal/room	onset 55.0, peak 63.0
water/normal/70	onset 84.6, inflection 90.0
hydraulic fluid/normal/70	N/A (data not conclusive)
fuel additive/normal/70	onset 84.1, peak 119.2
water/normal/85	onset 87.5, inflection 91.3
hydraulic/normal/85	onset 76.5, peak 110.8
fuel additive/normal/85	onset 65.1, peak 98.6

* There is a potential transition zone around 50 deg. C, but it is not as significant as that in the high 70 deg. C-mid 80 deg. C area. This specimen treated in water was expected to have a transition zone lower than for the case of control specimen at the same temperature, but this did not seem to occur.

The DSC tests indicated that the control adhesive specimens with normal cure had a transition point (possibly the T_g itself) ~40 deg. C below the published service temperature of 107 deg. C, Table 1. Baseline adhesive specimens prepared with normal cure and then tested at 70 deg. C and at 85 deg. C had a significantly higher T_g, which was close to the value for the post-cured specimens. This behavior mirrors the hardness values of these specimens. [31] However, the transition behavior occurs below the service temperature also in these conditions. Water treatment causes a small decrease of the transition range of the epoxy adhesive at 70 deg. C and at 85 deg. C with respect to the control at those temperatures. A decrease of T_g, is consistent with the plasticization of epoxy reported in the literature (e.g. [52] for woven composites).

Experimental results are somewhat ambiguous for treatment in water at room temperature (Table 1). Immersion in fuel additive or in hydraulic fluid at 70 deg. C or at 85 deg. C caused a lowered transition point/T_g, while the hydraulic fluid treatment at room temperature led to an increase of T_g, possibly due to a chemical reaction that triggered more polymer cross-linking and anti-plasticization. DSC data for the conditioned carbon/epoxy will be reported in future work. The T_g for the neat epoxy specimens (Proset® 117 LV/237) at room temperature was in the 60 deg.-70 deg. range (onset 63.4 deg. C, peak 77.5 deg. C).

Thermogravimetric results for selected control samples of the structural epoxy-based adhesive material were consistent among each other, with similarly shaped profiles. Only ~5-10% mass loss occurred up to 350 deg. C, followed by rapid mass loss between 350 deg. C and 500 deg. C and by a small additional mass loss (~2-3%) between 500 deg. C and 900 deg. C. Residual mass for the baseline ranged between 12% and 18%. Control samples, which were post-cured prior to the TGA runs, had a higher residual mass, mirroring the improved mechanical performance of those specimens with respect to those that were not post-cured.

Adhesive exposed to water or jet fuel, even for long periods of time and at high temperatures (e.g. 85 deg. C for 1,250 hrs), showed little variation in the TGA analysis, as seen in Figure 7. Profiles were unchanged, with a few percent of additional of total mass loss, indicating minor adsorption of the solvent, but no significant chemical or structural changes to the material. The mass gain from the gravimetric tests was approximately 8% for specimens immersed in water and jet-fuel.

Additional TGA scans of the structural adhesive treated in hydraulic fluid showed that there was no effect when the treatment took place at room temperature, consistently with the gravimetric data, which showed no mass uptake of hydraulic fluid at that temperature, Figures 7-8. The scans at 70 deg. C and 85 deg. C demonstrate that mass loss increases with increasing temperature. Significant mass loss of the hydraulic-fluid-exposed material occurs at a very low temperature: 40% loss at only 200 deg. C - a temperature, at which virtually none of the neat hydraulic fluid volatilizes, Figure 8.

TGA scans of the structural epoxy-based adhesives exposed to the fuel additive showed that this fluid by itself (no combination with jet fuel) had more impact than the hydraulic fluid, Figures 7, 9. Mass loss rate at the higher temperatures was somewhat greater but, more significantly, this fluid altered the structural epoxy-based adhesive already at room temperature. The 70 deg. C scans demonstrate that the effect of the solvent is reduced with post-curing of the epoxy-based adhesive.

Regarding the carbon/epoxy specimens, TGA plots indicated an effect for anti-icing additive treatments corresponding to approximately 10% more mass loss, but only at room temperature, Figure 10. This was consistent with the hardness changes, which were more significant at room temperature than at 70 deg. C. [31]

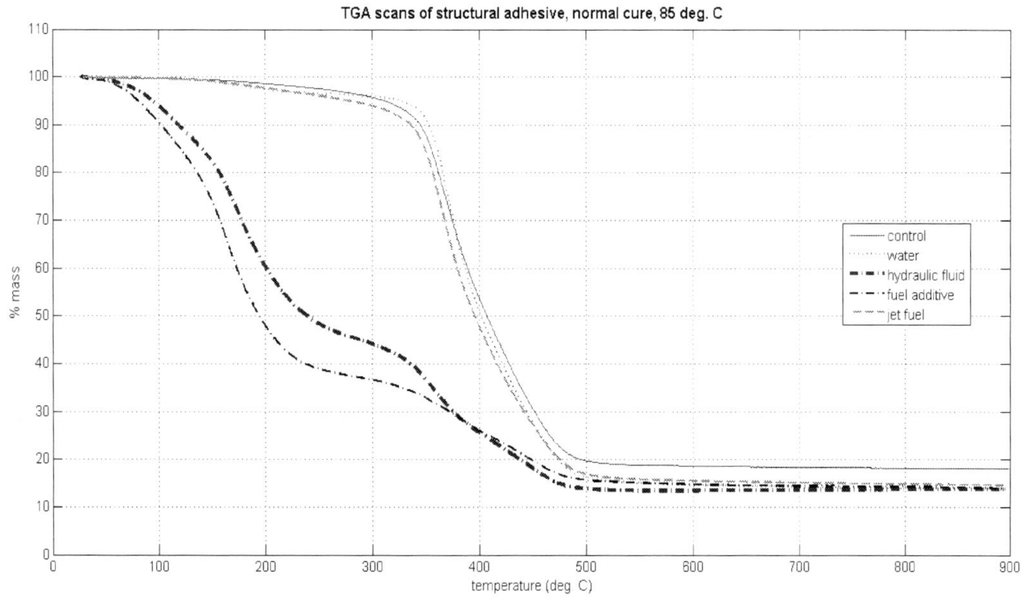

Figure 7. TGA scans of adhesive control and conditioned specimens, at 85 deg. C.

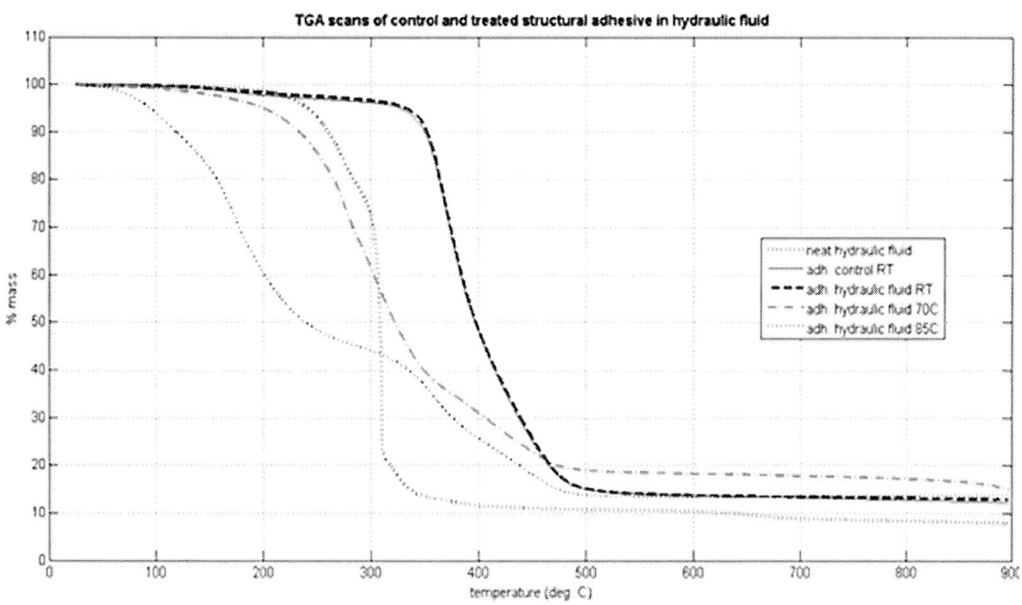

Figure 8. TGA scans of neat hydraulic fluid, control adhesive specimens, and conditioned adhesive specimens in hydraulic fluid, at room temperature, 70 deg. C and 85 deg. C (no post-curing).

In summary, thermogravimetric analysis and differential scanning calorimetry of the structural adhesive under consideration, exposed to the extreme conditions of full immersion in fuel additive or in hydraulic fluid, reveal that these exposures can drastically accelerate the volatilization of the structural epoxy-based adhesive.

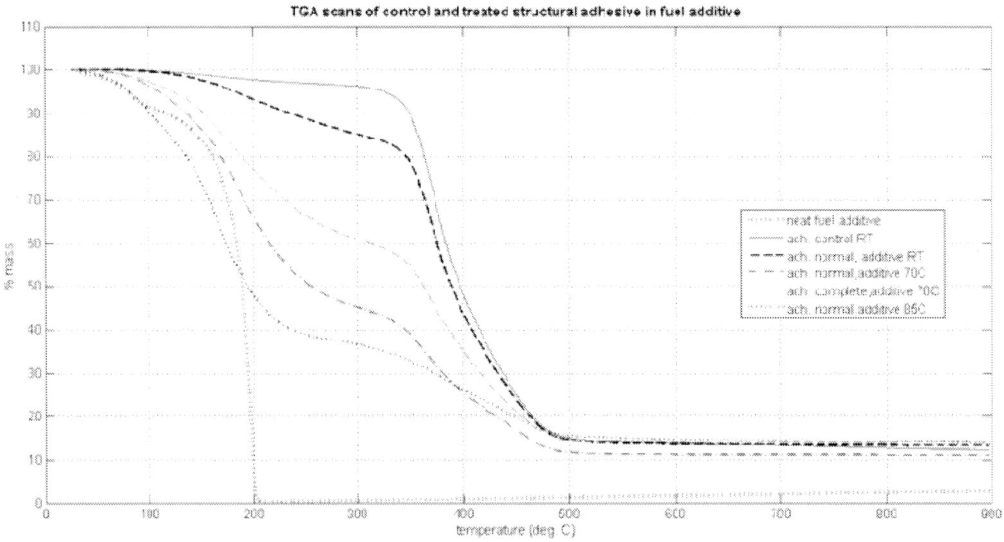

Figure 9. TGA scans of neat fuel additive, adhesive control specimens, and adhesive specimens treated in fuel additive. 'Normal' and 'complete' indicate respectively no post-curing and post-curing.

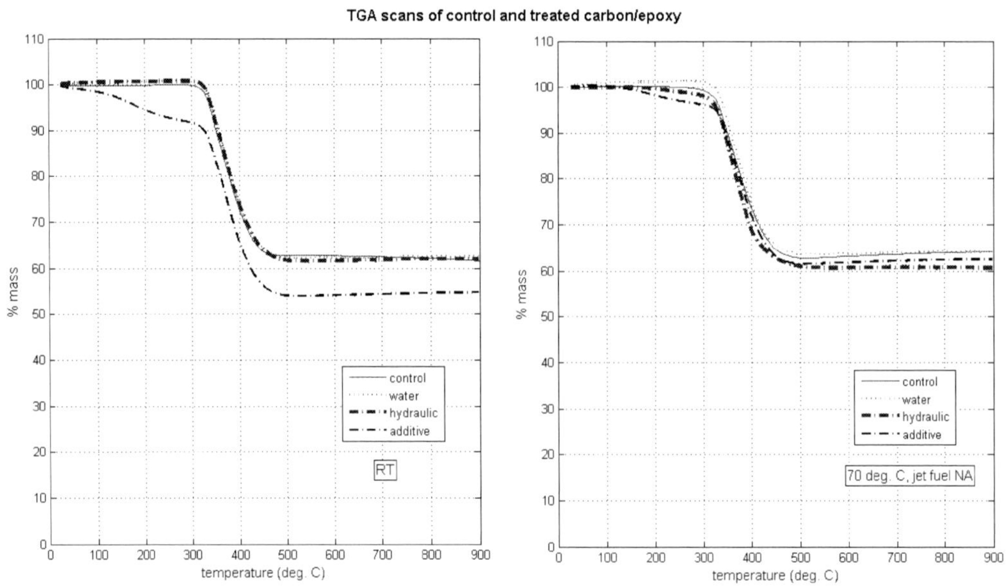

Figure 10. TGA scans of carbon/epoxy control and conditioned specimens, at room temperature ('RT') and 70 deg. C. Jet fuel-treated specimenswere not available.

There is significant mass loss at temperatures as low as 100 deg. C-200 deg. C versus more than 350 deg. C for unexposed material. This accelerated volatilization could potentially generate more combustible gas during the early stages of a composite fire incident, resulting in an increased risk.Additional work from our group will report the flammability characteristics of selected specimens frommicroscale combustion calorimetry tests (namely, specific heat release rate, heat release temperature, pyrolysis residue). [53]

CONCLUSION

This chapter delivers results of a study onstructural epoxy-based and carbon/epoxy, affected by many levels of mechanical and chemical degradation because of exposure to an aggressive environment (fluids typical of aerospace operations). Preliminary results on single lap joints highlighted the need to focus on the individual components of the joints. The project included gravimetric tests and hardness measurements analyzed with the help of statistics, [31], the adoption of two simple one-dimensional diffusion models able to fit gravimetric data under certain circumstances (water, fuel additive, see [50]). Moreover, TGA and DSC tests show drastically altered behavior for adhesive exposed to fuel additive or hydraulic fluid. Namely, mass loss occurred at much lower temperatures (100 deg. C-200 deg. C) than the unexposed samples (350 deg. C). Carbon/epoxy behavior appeared to have a small (~10%) change only for fuel additive treatment, and only at room temperatures. These TGA results parallel those of the gravimetric tests and hardness measurements.

Although simplified laboratory configurations were chosen (e.g. full immersion, no combination of fluids, no coatings, no prior thermo-mechanical fatigue), this research effort emphasizes that structural epoxy-based adhesive is greatly vulnerable to hydraulic fluid and anti-icing additive. The outcome on damage due to anti-icing additive is likely amplified by the lack of mixing with jet fuel. It is possible that the outcome in lap joints may be different, due to the location and different vulnerability of the adhesive in the joint, especially in the presence of commercial sealants.Additional tests on lap joints will follow at a later date.

This work indicates that the exposure to these fluids may have dire consequences on the mechanical performance and thermal stability of the adhesive in normal operations. It is recommended topay much attention to seal the exposed surfaces of these materials, and torepairany leaks from hydraulic fluid and jet fuel lines.

ACKNOWLEDGMENTS

The authors thank Charles Winkelmann and Yoshino Sugita (Advanced Composites Research, Engineering and Science, UC Davis) for manufacturing the specimens and collecting the gravimetric tests upon which this study is based. Also, the authors gratefully acknowledge the insightful commentsofDr. Robert Jensen, Materials Division, US Army Research Laboratory, as well as thefollow-up microcalorimetry work of Dr. Brent Pickett, US Air Force Research Laboratory (AFRL/RXQE).

This work was partially funded by the National Science Foundation (grant CMMI-062196 and CAREER grant CMMI-0642814) to author La Saponara.

The authors are grateful to AFRL/RXQE for supporting authors Campbell, Sullivan and Dierdorf, and funding the DSC and TGA tests.

REFERENCES

[1] American Composites Manufacturers Association (2003). Global FRP Use in Bridge Applications. http://www.mdacomposites.org/mda/bridge_statistics.htm.

[2] Bakis, C.E.; Bank, L. C.; Brown, V. L.; Cosenza, E.; Davalos, J. F.; Lesko, J. J.; Machida, A.; Rizkhalla, S. H.; Trantafillou, T. C. *J. Compos Constr*. 2002, 6, 73-87.

[3] Cheng, L.; Karbhari, V.M., *Prog. Struct Eng. Mat*. 2006, 8, 143-154.

[4] Bocciarelli, M.; Colombi, P.; Fava, G.; Poggi, C. *Compos Struct*. 2009, 87, 334-343.

[5] Hollaway, L. C. *Constr. Build Mater*. 2010, 24, 2419-2445.

[6] Durability of composites for civil structural applications; Kharbari, V. M.; Woodhead Publishing Limited: Cambridge, England, 2007.

[7] Red, C. C. (2008). Wind turbine blades: Big and getting bigger. Composite Technology. http://www.compositesworld.com/articles/wind-turbine-blades-big-and-getting-bigger.

[8] American Wind Energy Association (2009). Wind Power Outlook 2009. http://www.awea.org/pubs/documents/Outlook_2009.pdf

[9] Engineering Talk (2005). Composite catches train for safety requirements. http://www.engineeringtalk.com/news/gad/gad176.html

[10] Marsh, G. *Reinforced Plastics magazine*. 2004, 48, 26-28.

[11] Niu, M. C. Y. Composite Airframe Structures –Practical Design Information and Data; Adaso/Adastra Engineering Center: Granada Hill, CA, 2000; 3rd edition.

[12] Hart-Smith, L. J. In *Bonded Repair of Aircraft Structures;* Baker, A. A.; Jones, R; Martinus Nijhoff Publishers: Dordrecht, The Netherlands, 1988; 31-47.

[13] Hart-Smith, L. J., J *Adhesion*. 2006, 82, 181-214.

[14] da Silva, L. F. M.; Adams, R. D. *Int. J. Adhes. Adhes*. 2007, 27, 362-379.

[15] Armstrong, K. B.; Bevan, L. G.; Cole, W. G II. *Care and Repair of Advanced Composites;* 2nd edition; SAE International: Warrendale, PA, 2005.

[16] Hashim, S. A.; Knox, E. M. J. Adhesion. 2004, 80, 569-583.

[17] Kim, S. S.; Yu, H. N.; Lee D. G. *Compos. Sci. Technol*. 2007, 67, 3417-3424.

[18] Osnes, H.; McGeorge, D. *Compos Part B-Eng*. 2009, 40, 29-40.

[19] Keller, M.; Schollmayer, M. *Compos. Struct.* 2009, 87, 232-241.

[20] Silva, M. A. G.; Biscaia, H. *Compos. Struct.* 2008, 85, 164-174.

[21] Ashcroft, I. A.; Crocombe, A.D. In *Modeling of Adhesively Bonded Joints*; da Silva, L. F. M.; Öchsner A. Springer-Verlag: Berlin, Heidelberg, Germany, 2008; 183-223.

[22] Bonded Repair of Aircraft Structures; Baker, A. A.; Jones, R; Martinus Nijhoff Publishers: Dordrecht, The Netherlands, 1988.

[23] Modeling of Adhesively Bonded Joints; da Silva, L. F. M.; Öchsner A. Springer-Verlag: Berlin, Heidelberg, Germany, 2008.

[24] Fatigue in Composites –Science and technology of the fatigue response of fibre-reinforced plastics; Harris, B.; Woodhead Publishing Limited: Cambridge, England, 2003.

[25] Bar-Cohen, Y. (2000). Emerging NDE Technologies and Challenges at the Beginning of the 3rd Millennium, Part II, NDT.net, 5, http://www.ndt.net/article/v05n02/barcohen/barcohen.htm

[26] Mouritz, A. P.; Gibson, A.G. *Fire Properties of Polymer Composite Materials*; Springer: Dordrecht, The Netherlands, 2006.

[27] Lyon, R. E.; Janssens, M. L.; Polymer Flammability; Federal Aviation Administration: FAA-05/14 (DOT/FAA/AR-05/14), USA, 2005, www.fire.tc.faa.gov/pdf/05-14.pdf

[28] Quintiere, J. G.; Walters, R. N.; Crowley, S. Flammability Properties of Aircraft Carbon-Fiber Structural Composite. US Department of Transportation: DOT/FAAR/ AR-07/57, USA, 2007, www.fire.tc.faa.gov/pdf/07-57.pdf

[29] Gandhi, S.; Lyon, R. E.; Health Hazards of Combustion Products From Aircraft Composite Materials. US Department of Transportation: DOT/FAAR/AR-98/34, USA, 1998. www.tc.faa.gov/its/worldpac/techrpt/ar98-34.pdf

[30] Asaro, R. J.; Krysl, P.; Zhu, B.; Ramroth, W. T. *Compos. Struct.* 2005, 68, 399-408.

[31] Sugita, Y.; Winkelmann, C.; La Saponara, V. *Compos. Sci. Technol.* 2010, 70, 829–39.

[32] Dao, B.; Hodgkin, J.; Krstina, J.; Mardel, J.; Tian, W. J. *Appl. Polym. Sci.* 2006, 102, 3321-3232.

[33] National Transportation Safety Board, Safety Recommendation A-06-27 and -28 (March 24, 2006), USA.

[34] Henkel, Structural Adhesive Pastes for Aerospace MRO Applications (2010). http://www.henkel.com/com/content_data/131889_Structural_Adhesive_Pastes_for_MRO.pdf

[35] Henkel, Technical Data Sheet for Hysol EA 9360 (2010). http://www.henkelna.com/cps/rde/xchg/SID-0AC83309-FC2401D9/henkel_us/hs.xsl/full-product-list7932.htm?iname=Hysol+EA+9360andcountryCode=usandBU=industrialandparentredDotUID=productfinderandredDotUID=0000000ISNandparam1=technical.

[36] Rider, A.; Yeo, E. The Chemical Resistance of Epoxy Adhesive Joint Exposed to Aviation Fuel and its Additives. Australian Government Department of Defense. 2005, DSTO-TR-1650. http://dspace.dsto.defence.gov.au/dspace/handle/1947/4067

[37] Curliss, D. B. The effect of jet fuel exposure on advanced aerospace composites I: thermal and chemical analyses. Materials Directorate, Wright-Patterson Air Force Base: Report WL-TR-91-4017, 1991. http://oai.dtic.mil/oai/oai?verb=getRecordandmetadataPrefix=htmlandidentifier=ADA246559.

[38] Bagley, E., Long, F.A. *J. Am. Chem. Soc.* 1955, 77, 2172-2178.

[39] Astarita, G., Nicolais, L. *Pure Appl. Chem.* 1983, 55, 727-736.

[40] Mensitieri, G., Lavorgna, M., Musto, P., Ragosta, G. *Polymer.* 2006, 47, 8326-8336.

[41] Helbling, C. S., Kharbari, V. M. *J. Reinf. Plast Comp.* 2008, 27, 613-638.

[42] Crank J. *The Mathematics of Diffusion*; Oxford University Press: New York, NY, 1975, 2nd edition.

[43] Carter, H. G., Kibler, K. G. *J. Compos. Mater.* 1978, 12, 118-131.

[44] Shen, C.-H.; Springer, G. S. *J. Compos. Mater.* 1976, 10, 2-20.

[45] Popineau, S.; Rondeau-Mouro, C.; Sulpice-Gaillet, C. ; Shanahan, M. E. R. *Polymer.* 2005, 46, 10733-10740.

[46] Perreux, D.; Suri, C. *Compos. Sc.i Technol.* 1997, 57, 1403-1413.

[47] Lee, M. C.; Peppas, N. A. *J. Appl. Polym Sci.* 1993, 47, 1349-1359.

[48] LaPlante G.; Ouriadov, A. V.; Lee-Sullivan, P.; Balcom, B. J. *J. Appl. Polym. Sci.* 2008, 109, 1350-1359.

[49] Ferreira, J. M., Pires, J. T. B.; Costa, J. D.; Errajhi, O. A.; Richardson, M. *Compos.Struct.* 2007, 78, 397-401.

[50] La Saponara, V. *Compos Struct.* January 2011. Revised paper under review.
[51] Canfield, R. (2001). Central differences MATLAB routine. http://www.mathworks.com/matlabcentral/fileexchange/12-centraldiff-m
[52] Abot, J. L.; Yasmin, A.; Daniel, I. M. *J. Reinf. Plast. Comp.* 2005, 24, 195-207.
[53] Campbell, R.; Pickett, B.; La Saponara, V.; Dierdorf, D. Thermal characterization and flammability of structural adhesive and carbon/epoxy composites with environmental and chemical degradation. Submitted to *J. Adhes. Sci. Technol.*, January 2011.

In: Adhesive Properties in Nanomaterials, Composites ... ISBN: 978-1-61209-268-3
Editors: K. A. Wilkinson and D. A. Ordonez

Chapter 3

PULSED HIGH- AND LOW-ENERGETIC FILM GROWTH ON THERMOPLASTIC POLYURETHANE BY PULSED LASER DEPOSITION AT ROOM TEMPERATURE

J. M. Lackner[1,*], W. Waldhauser[1], R. Major[2], and B. Major[2]

[1]Joanneum Research Forschungsgesellschaft mbH, Institute of Surface Technologies and Photonics, Niklasdorf, Austria

[2]Polish Academy of Sciences, Institute of Metallurgy and Materials Sciences, Krakow, Poland

ABSTRACT

The coating of polymer materials by protecting and functional films requires high efforts in the development of coating techniques due to the very different mechanical and thermal properties of polymer substrates and metal or ceramic films. The film has to fulfil both high adhesion and optimized microstructure to prevent failing in its application. The current work describes a new vacuum coating technique for polymer materials with inorganic films – the Pulsed Laser Deposition (PLD) process, characterized by a high-energetic pulsed plasma and the easy possibility to room temperature deposition (RT-PLD). Thus, pseudodiffusion interfaces were found due to the high-energetic particle bombardment during PLD coating. Additionally, changes of the polyurethane chemical binding are evident, like the transition from C=O to C–O–R binding, in which titanium atoms could act as new binding partners to the O species. Although very high film adhesion can be guaranteed in the PLD by the formation of pseudodiffusion interfaces, preventing the well-known buckling phenomenon, high film stresses result in plastic deformation of the soft polymer surface and the formation of wrinkles. The reasons and effects of wrinkling – even starting in growing films – on the film behaviour are described in this work, based on both practical investigations, using transmission electron

* Joanneum Research Forschungsgesellschaft mbH, Institute of Surface Technologies and Photonics, Leobner Strasse 94, A-8712 Niklasdorf, Austria,Tel.: +43 316 876 2305, Fax: +43 316 876 2310, Email: juergen.lackner@joanneum.at

microscopy, X-ray diffraction and atomic force microscopy, and theoretical finite element modelling.

1. Introduction

Surface modifications of polymers by coating allow new applications by e.g. adapted wear protection, chemical reactivity, biocompatibility, electrical, and thermal conductivity. Adhesion of the coatings is a major factor in their performance. The interfacial region is of special interest, since it is the weak point of the polymer-coating compound. The practical adhesion is the relevant value in all investigations, because it gives hints towards the usability of thin films. [1] Substantial changes in the chemical functionality, surface texture, wettability, and bondability of polymer surfaces can be achieved by plasma treatment (ion bombardment) of these materials [2, 3] allowing improved adhesion of vapour deposited coatings. Many theories proposed mechanisms for increasing the adhesion of these coatings, like the introduction of new reactive chemical species, the elimination of weak boundary layers, actual chemical changes such as oxidation, and increased surface roughness due to pitting. Recently it turned out, that the incorporation of new reactive chemical species into the polymer surface, e.g. to form polar functional groups, seems to be of prime importance [4] in the coating of polymers with metals and metal compounds (oxides, nitrides, carbides). Reactive surface functional groups, especially oxygen-containing groups (polar groups like C-O or C=O) play the major role in the bonding of the first monolayers due to their high reactivity with metal atoms. [5-9] Such functional groups bonded to metal atoms were also found in the interface zone between pulsed laser deposited (PLD) coatings and polymers. [10]

In the PLD technique each laser pulse leads by the interaction with the (metallic) target to the formation of a high intensity plasma. Due to the specific ablation mechanisms, the plasma pulses contain two fractions of particles with different energy distribution: The first pulse contains very high energetic ions with energies between 100 and 1000 eV, the second lower energetic species between 10 and 50 eV. [11] The vaporized material consisting of atoms, ions and atomic clusters is subsequently deposited onto the substrate surface. Due to the faster expansion of the higher energetic flux reaching the substrate surface earlier than the lower energetic flux, these impinging ions can remove (sputter) loose bonded atoms from the surface, can penetrate into the polymer surface, can cut chemical bonds in the polymer, but also can be resputtered. [12] Thus, the formation of pseudodiffusion interfaces can be expected [10], which are very advantageous for the high adhesion of coatings. In contrast, most of the other, competitive vacuum coating techniques (e.g. magnetron sputtering, thermal evaporation, chemical vapour deposition) struggle with low particle energy in the thermal range (several eV). Such particles are not able to penetrate a substrate (polymer) surface deeply preventing the formation of well-adherent, coatings with pseudodiffusion interfaces. Only abrupt interfaces can be found, which are anchored to the outermost atomic layers of the substrate by chemical bonds or via roughness tips.

2. Mechanisms of Debonding of Coatings on Polymers

The adhesion of the first deposited atoms of the metal coating on the polymer surface and the type of interface formation strongly influences its performance. However, the subsequent growth behaviour of the film on the very soft, easy elastically deformable polymer surface has consequences on the mechanical behaviour of the compound too. It is well known, that coatings deposited at low temperatures by various PVD techniques possess high growth stresses, because bulk diffusion is disabled. However, for the coating of polymer materials these low substrate temperatures are decisive. Thus, growth defects, like trapped contaminating atoms (e.g. from the deposition atmosphere), atoms on wrong positions in the crystal lattice, or lattice vacancies, cannot migrate from the lattice inside the grown crystals to the grain boundaries [30], where because of the amorphous state the packing density is lower than in the crystal lattice leading to lower stress fields. Compressive stresses are increasing with increasing solute concentration c. For an elastically compressible film the compressive stresses σ can be estimated by (eq. 1):

$$\sigma = -\frac{E_1}{3(1-\nu)}(\frac{\Delta V}{V}) = -\frac{E_1 \nu_p c}{3(1-\nu)}, \qquad \text{(eq. 1)}$$

in which ν_p is the partial molar volume of the solute, E the Young's modulus, ν the Poisson ration and $\Delta V/V$ the theoretically volume expansion of the coating. This relation let expect higher compressive stresses in the lattice as higher not only the concentration and the partial molar volume of the solute is, but also as higher the elastic modulus of the layer is.

To minimize the stress level, the structure tends to deform. Deformation in PVD coatings deposited at low temperatures is mainly possible in the regions of lower packing density between the (micro)columns. [31] These (micro)columns are developing due to the distinct growth conditions. Their growth starts from initial nuclei; subsequently selective and preferred growth of some of these nuclei results in the formation of the observed columnar growth structures (zone-1 and zone-T structures of Thornton's structure zone model). Increased density of the coating decreases the deformability by densification of the intercolumnar regions. Generally, increased density of the coatings is reached by higher growth temperatures or by activating the atom mobility (surface diffusion) by high-energetic (ion) bombardment of the growing film. [32] As mentioned above, the pulsed plasma of the PLD technique allows the high-energetic film growth conditions, necessary for the formation of dense films even at room temperature. Although increased adhesion and densification of the microstructure are the main advantages of the ion bombardment, this technique results in an increased number of lattice defects too. Mainly ions and atoms (e.g. Ar from the process gas) penetrate the film during growth due to their high energy. Thus, the stress level is increased in these PVD coatings. [33, 34] The stresses in PLD hard coatings (e.g. TiN on steel substrates) can reach several GPa, in hard coatings) on steel substrates. [35, 36]

Stress relaxation in high adhesive, dense films is only possibly by (elastic and plastic) deformation of the whole system – the substrate and the film. Thus, the stress level in the film strongly depends on the mechanical properties of the film, substrate surface and interface region. Macroscopically, stress and deformation are closely connected by relations for the elastic (described by Hook's law and the Young's modulus) and plastic (described by the yield strength, work hardening and tensile strength) range, resp.. These two ranges are closely connected to elastic deformation of the lattice (lattice strain) and plastic flow due to dislocation movement.

2.1. Wrinkling of Coatings

For reduction of the overall strain energy in a compressive stressed coating, wrinkling has been suggested as a mode of common deformation of the substrate (surface) and the film. [37] Wrinkling is described as the formation of sinusoidal buckles on the surface, which do not loose their adhesion to the substrate distinguishing it from buckling. Thus, the main deformation mechanisms in wrinkling are bending of the substrate or of the near-surface region of the substrate together with the covering coating. The mechanical problem of winkling – the mechanical instability of a thin film on a flexible substrate – can be compared to the instability problem of an elastic layer covering an elastic medium [38-40] described by the Euler solution of the mechanical instability of a plate under an unaxial compression force. [24] In this model, all appearing stresses (both in the coating and in the substrate) have only biaxial character with both components parallel to the surface – deformation is impossible in these directions. In contrast, deformation is allowed in the direction normal to the surface. For decreasing the complexity of this instability problem, in the following the model is simplified to two dimensions and deformation is impossible in x-direction. The result of releasing the compressive stresses (if they have reached the critical value for starting of wrinkling [42]) is the formation of a wavy structure at the surface. The distance between the wrinkles (the wavelength of the wavy structure topography) can be expressed by [42] as (eq. 2)

$$\lambda = 2\pi h \sqrt[3]{\frac{(1-\nu^2)E_1}{3(1-\nu_1^2)E}} \qquad \text{(eq. 2)}$$

and, thus, strongly depends on the film thickness h and the Young's modulus and Poisson ratio of the coating (E_l and ν_l) and the substrate (E and ν), resp.. As higher the film thickness is, as higher is the wavelength and as lower is the number of wrinkles. Additionally, the wavelength of wrinkling is increased by higher stiffness of the coating and higher flexibility of the substrate. The stress level in the coating is strongly dependent on the elastic properties in the substrate. Stress relaxation in the coating is reached by bending of the whole compound – the attained radius of the curvature R is related to the compressive residual stress σ in the coating and can be approximated by the Stoney's equation (eq. 3) [43]

$$\sigma = \frac{-E}{6(1-\nu\ )}\frac{s^2}{hR}, \qquad \text{(eq. 3)}$$

in which E, ν and s are the Young's modulus, Poisson's ratio and thickness of the substrate.

As easier the substrate can be deformed, as higher is the tendency to wrinkling and as lower is the stress level in the film. In the case of coating metals or semiconductors, the stress relaxation is very small due to their difficult deformation.

2.2. Buckling of Coatings

If the in-plane stress level in the film exceeds a critical value, crack formation at the interface between the film and the substrate occurs, leading to local or global failure of the film. Hence, the nucleation and growth of individual blisters over initially de-bonded patches are found. [44-54] Spontaneously buckling patterns of various dimensions and structures develop, depending on the internal stresses and film thickness. The buckling phenomenon can be simplified described using the adhesion energy.

If the released elastic energy is equal or larger than the energy of adhesion, buckling is energetically possible. [55] By introducing γ as the adhesion energy by unit area, the corresponding critical stress for buckling σ_c (eq. 4)

$$|\sigma_{cr}| = \pm|\sigma| + \sqrt{\frac{2\gamma E}{D(1-\nu^2)}}, \qquad \text{(eq. 4)}$$

in which the positive sign of the stress σ in the film (induced by solute molecules or atoms) corresponds to tensile initial stresses ($\sigma > 0$) and the negative sign to compressive initial stresses.

Buckles occur in different topographical structures, such as typical circular blisters (hemispherical caps, worm-like structures, or straight-sided winkles). [44-54] The different shapes of buckles might be correlated with the elastic and bending energy stored in the lifted area of the film. [56] Especially corners and sharp edges in the buckles will raise this energy contribution. The elastic energy contribution scales with the film thickness and might be negligible in thin films. Therefore, the film buckles by forming hemispherical caps formed all over the sample surface, which efficiently release in-plane stresses in both in-plane directions. For thicker films, the bending energy cannot be neglected any more: Hence, this energy contribution elongates buckles in a certain direction and the buckle keeps that direction as long as possible. However, such an elongated, straight-sided buckle can only release in-plane stresses in one in-plane direction. Therefore, it is forced to include an edge after a certain length that directs the buckle more in the perpendicular direction (worm-like structure, telephone-cord buckling). By this, the perpendicular component of the in-plane stresses can also be released.

2.3. Wrinkling and Buckling on the Nanometer Scale

Both phenomena were illustratively shown by Mylvaganam et al. [57] in structural calculations – releasing biaxial compressive stresses of the film to the substrate – on the atomic scale (Fig. 1). These calculations, based on forces between the atoms, show for thin diamond coatings on (100) orientated single-crystal silicon substrates [57] two distinct characteristics:

(1) The formation of a clustered surface topography with a quite regular arrangement of the wrinkles (buckles) (Fig. 1b) allows a relaxation of the compressive stress in the coating by its deformation. The "volume" of the coating is increasing, clearing space for the expansion of the grown lattice. However, full stress relaxation is only possible in some regions. The formation of the wavy surface creates in the curved areas again tensile and compressive stresses. Using the example of the top of such a wave, the upper part of the coating is tensile stressed, the lower part compressive stressed (and vice versa for a valley in such a wave structure). [37] Consequences of this behaviour are shown in the following.

(2) The formation of a disordered surface region of the silicon substrate (Fig. 1a) is found due to the growth stresses developing in the coating. Loosing the crystal structure in the near-surface region of the substrate results in lower packing density of the atoms and, hence, in a worse bond allowing easier deformability by higher elasticity and lower stress levels for plastic flow mechanisms (weak-boundary layer). In Fig. 1a the partly delamination of the coating in the near-surface region of the substrate was found (exceeding the fracture stress of the interface).

These theoretical considerations of the effects of growth stresses on the coatings performance reveal that mainly the mechanical properties of the substrate, coating and the interfacial region control the deformation. Reasonable, the coatings behaviour like roughness, topography, but also applied properties like the electrical conductivity or friction and wear significantly depend on the growth stresses. As shown on the given example of different residual stress levels on substrate materials with different elastic modulus, it is obvious, that the largest consequences of coating growth stresses are found on the materials of the lowest elastic moduli – polymers and organic compounds.

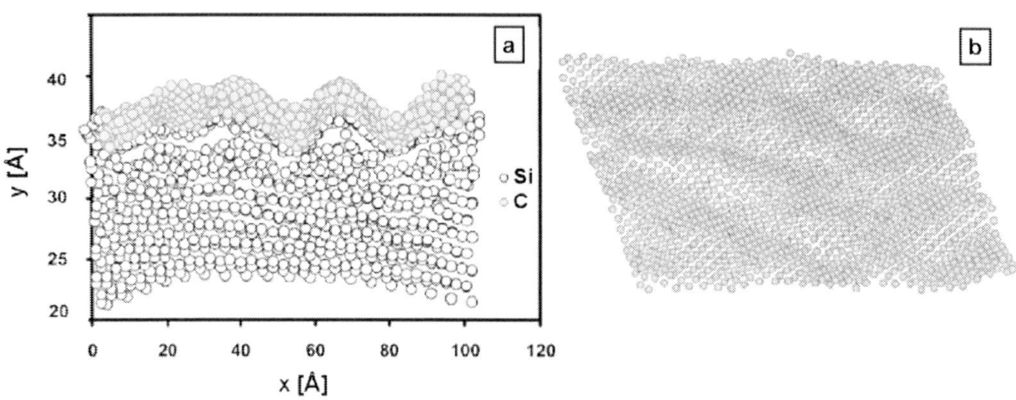

Figure 1. (a) Calculated cross-section of a carbon film on a silicon substrate and (b) topography of the film after releasing the biaxial stresses (after [57]).

It is clear, that the model used for calculations in Fig. 1 is widely simplified, because as "static" assumption the stress was released in the coating without any subsequent growth after this deformation. In a growing coating the stresses increase, if the thickness of the coating increases. Studying the behaviour of the substrate under such conditions seems to be interesting, but this is scarcely possible by applying substrate materials of low propensity to deformation (high elastic modulus, high yield strength) like metals, semiconductors or ceramics. In contrast, using high-elasticity materials (polymers) as substrates allow an easier opportunity for investigating these phenomena. Additionally, these investigations enable the understanding of the processes occurring in the coating of polymers by plasma deposition techniques as well as giving hints for improving such material compounds.

3. Experimental Details

3.1. Film Deposition

In the current work, two different deposition techniques were applied to show the differences between low- and high-energetic film growth on polymers: The conventional balanced direct current (DC) magnetron sputtering of pure aluminum films from an Al target in argon atmosphere was used for low-energetic film growth at room temperature. To show the differences occurring in high-energetic deposition, the PLD was used for depositing titanium based films by ablating pure titanium (99.9 % Ti) targets with a pulsed Nd:YAG laser system at room temperature (RT-PLD), too. This industrially-scaled system provides four beams of 1064 nm wavelength with 0.6 J pulse energy and 10 ns pulse duration at a repetition rate of 50 Hz. Argon was used as inert gas for pure titanium coating, nitrogen as reactive gas for titanium nitride deposition. Additionally, magnesium, nickel, and chromium targets were used for the comparative deposition of these pure metal films in pure argon atmosphere.

The substrate material used was a thermoplastic polyurethane (TPU) casting resin of 2 mm thickness and an aromatic-based polyurethane structure. Coating of this special type of TPU by traditional PVD coating techniques is impossible due to its high elasticity (tensile modulus at 100 % elongation: 6.5 to 20 MPa, ultimate elongation: 300 to 550 %, ultimate tensile strength: 35 to 45 MPa), the low hardness (80 Shore A to 75 Shore D), the low Vicat softening temperature between 80 and 110 °C and the low glass transition temperature between -35 and -25 °C.

3.2. Film Characterization

X-ray photoelectron spectroscopy (XPS) was employed to investigate the chemical bonding in the films and the interfacial region using an Omicron Multiprobe system with a monochromized AlKα (1486.6 eV) X-ray beam and an EA 125 energy analyser. The resolution of this set-up is better than 0.3 eV, and the analysis took place at a pressure of 4×10^{-9} Pa. The spectrometer was operated in the fixed analyzer transmission mode. Neutralization of the non-conductive polymer surface was performed by an electron gun. All binding energies reported in this work were referenced to the binding energy of the carbon C1s peak at 285.0 eV. The detection sensitivity was about 1 mass %. For depth analyses an Omicron "ISE 10" sputter gun using 2 keV Ar^+ ions was used. The sputtering depth was determined by subsequent atomic force microscope imaging of the sputtering pit.

For verifying the XPS data Fourier-transformed infrared (FTIR) spectroscopy in attenuated total reflection (ATR) setup was applied with a Perkin Elmer Spectrum One system equipped with a quartz/halogen lamp and a deuterated triglycine sulphate detector (wavelength range: 350 – 7800 cm^{-1}). ZnSe was choosen as ATR crystal material, allowing the internal reflectance of the infrared light and, thus, the increase of the peak intensity in this setup. For ensuring full contact of the Ti coated PU surfaces (film thickness: 5 nm, 20 nm) to the ATR crystal, the sample (diameter ~ 5 mm) was pressed with a force of 10 N onto it.

The surface quality and structure of the coatings were inspected by light and scanning electron microscopy (SEM, Cambridge Instruments Stereoscan 360, acceleration voltage: 30 kV). The SEM was used for micro tensile testing too. Atomic force microscopy (AFM) was applied in tapping mode for quantifying the surface topography on a nanometer scale using a Nanoscope III (Digital Instruments) facility. The phase composition and texture analyses of the coatings were performed using a Philips PW 1710 X-ray diffractometer (XRD) and CoKα radiation. The pole figure measurements were performed on a four cycle goniometer equipment with a step size of 2.5° for the polar and azimuthal angle. Transmission electron microscopy (TEM) on a Philips CM20 (LaB_6) equipment operating at 200 kV was used for the microstructural investigations and growth structure analyses. The sample preparation procedure included mechanical cutting, slicing, polishing, and finally Ar^+ ion milling for reaching an end-thickness of approximately 50 nm. Electron micro-diffraction measurements were carried out for determining the phase composition of areas down to 40 nm diameter. For chemical analyses, the TEM was equipped with an energy dispersive spectroscopic (EDS) analyzer.

4. Structure and Chemical Binding of PLD Coatings on Polymers

TEM investigations, as shown in Fig. 2, revealed a full and dense covering of the wavy polyurethane surface by the about 50 nm thick titanium film. Although during TEM sample preparation the coated compound was highly strained (bending, shearing), cracking of the coating was scarcely found, revealing extremely high adhesion of the PLD film. The microstructure of the Ti film was found crystalline (compare the electron diffraction pattern in the insert of Fig. 2), the slight contrast differences in the bright field image let expect a very fine-grained micro-columnar microstructure. The same behaviour of film growth by PLD at room temperature was found for the coating of metal substrates and is unique for a room temperature coating technique (e.g. explained in [14]). Reasons therefore are the processes occurring during the impinging of pulsed, both high- and low-energetic plasma fluxes on the substrate (and growing film) surface. As described above, the high-energetic species is able to penetrate the substrate material. Thus, the formation of pseudodiffusion interfaces between the polymer substrate and the metal film can be expected. Comparing to the adhesion theories of Mattox [15], pseudodiffusion interfaces are characterized by the enforced incorporation of film atoms in the surface-near region of the substrate. To reveal such a pseudodiffusion layer formation for the PLD on polyurethane too, EDS analyses of carbon, nitrogen, oxygen, and titanium were performed in the high-resolution TEM. Although these images (Fig. 2) let easily distinguish between the film and the substrate material, the limit of detection of the aerial distribution of the atomic species was too high to find pseudodiffusion regions.

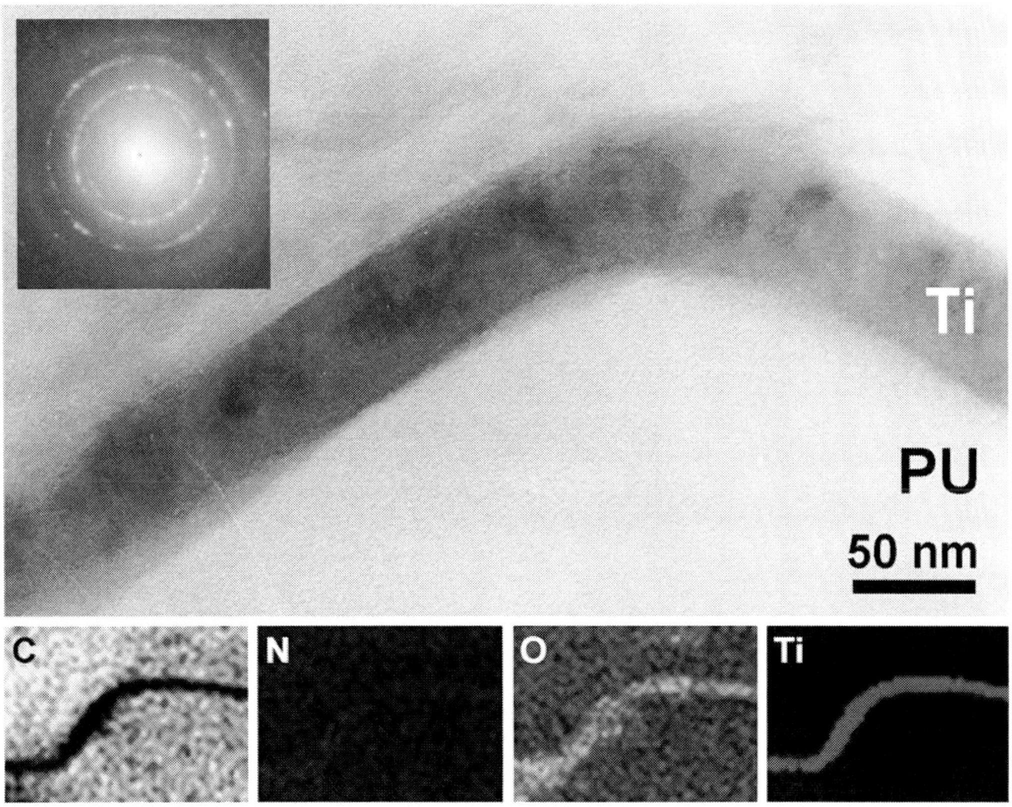

Figure 2. TEM images of the cross-section of an about 50 nm thick, room-temperature deposited PLD titanium film on a polyurethane (PU) substrate. (Upper row: bright field image and electron microdiffraction pattern; Lower row: EDS images of carbon, nitrogen, oxygen, and titanium species).

In a second step, XPS analyses were performed on an only 20 nm thick PLD Ti film. XPS allowed additionally to the detection of the chemical elements an analyses of the bonding structure. It was clear, that changes of the chemical bonding of the polymer chains could be expected, if high-energetic metal atoms are deposited and implanted in polymer materials. XPS depth analyses revealed an incorporation of laser ablated atoms up to 50 nm into the investigated polyurethane (Fig. 3). By using lower pass energy, remains of Ti species were found in up to 110 – 120 nm distance to the interface. The depth scans of the Ti, C, and N peaks revealed very clearly the two interesting zones – the grown film and the pseudodiffusion layer, which represents the transition to the bulk of the polyurethane substrate. On the surface scans the well-known contaminations (C, N, O) were evident, caused by the sample handling in ambient air atmosphere. Only short sputtering (~ 5 nm beneath the surface) allowed the measurement of the binding energy (representing the chemical binding) in the film. Compared to the surface scan, the contaminating carbon (bonded as C–OR) was drastically diminished, and the TiO_2 peak disappeared. In contrast, Ti–O binding with low oxygen content and Ti–N binding were dominating in the film. This incorporation of O and N was caused by both resputtering from the polyurethane surface (dissociation of the polymer molecules) and the deposition atmosphere. [16] Additionally, Ar species were found in the films, originating from the deposition atmosphere. After 20 nm sputtering (from the coating surface) the pseudodiffusion zone is reached, which was evident by the strongly changing spectra of the Ti and C peaks? Binding between Ti and O atoms was shifted to higher O contents (Ti_2O_3). The amount of C=O binding was found to be rather low, but C–OR binding was highly dominating. Comparing these findings with the structure of polyurethane – mainly C–O, C–N, and C=O bonds in the functional group, which binding was found in greater (sputter) depth beneath the surface – the highly dominating C–O–R, TiO_x ($x>1$), and nearly missing C=O peaks let expect the cutting of –C=O bonds and the formation of C–O–Ti and C–O–Ti–O–C groups in the pseudodiffusion layer.

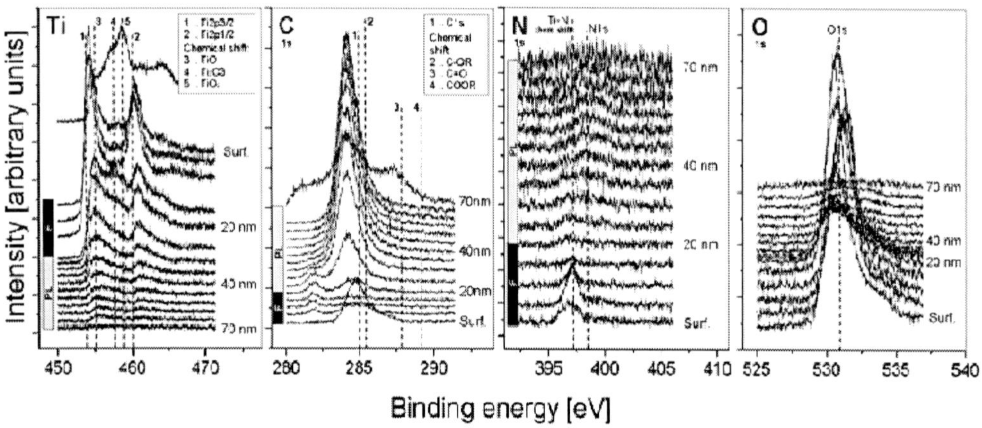

Figure 3. XPS spectra of the Ti, C, and N peaks taken from a depth analysis of an about 20 nm thick room temperature deposited PLD titanium film (labelled *F*) on a polyurethane substrate (*PL*: pseudodiffusion layer). Although neutralization of the non-conductive polymer surface was performed during the XPS measurements, small shifts (~ 1 eV to lower binding energies) are found for all depth scans excepting the scan at the surface.

In contrast, the binding between Ti and N (Ti–N) could replace H atoms or cut polymer molecule chains. As higher the sputtering depth was, as lower is the intensity of the Ti peak, but as higher is the intensity of the C peak. A totally missing Ti peak was found even after about 130 – 140 nm sputtering, presenting the total transition to the polyurethane bulk composition.

These XPS results are confirmed by FTIR-ATR spectroscopy investigations: The difference spectra (Fig. 4) show in the wavelength range between 3000 and 3750 cm^{-1} a strong decrease of the peaks for O-H group stretching and C-N binding. The first effect – decreased O-H group stretching – can be assigned to a transformation of O-H binding to O-R binding (R: e.g. Ti atoms), revealing the highly dominating C-O-R binding found in the XPS investigations. The second effect – decrease of C-N bonds – confirms the tendency to Ti-N binding by the dissociation of C-N bonds in the surface of the polymer.

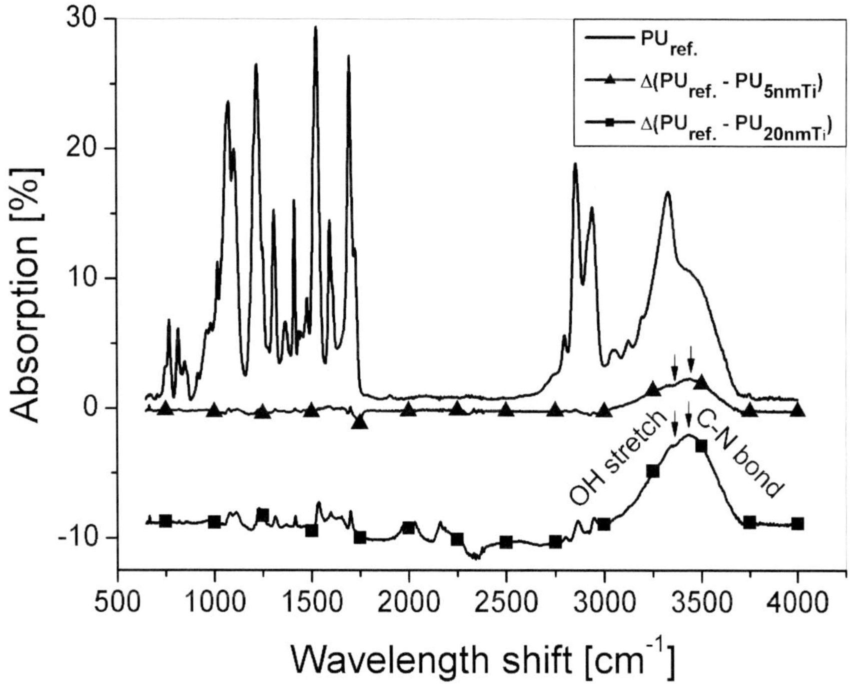

Figure 4. FTIR-ATR spectrum of the uncoated polyurethane ($PU_{ref.}$) and difference spectra between the uncoated and with 5 and 20 nm Ti coated polyurethane surfaces [$\Delta(PU_{ref.} - PU_{5nmTi})$, $\Delta(PU_{ref.} - PU_{20nmTi})$].

The high metal atom implantation depth (> 140 nm) for PLD Ti was revealed by works of Zaporjtchenko et al. (e.g. [17]), who investigated the high-energetic noble metal implantation in polymer surfaces. Both effects of the pulsed plasma conditions of the PLD technique – pseudodiffusion layer formation and chemical binding (chemisorption) – let expect increased adhesion on polymers, even on the atomic level, compared to other well established vacuum polymer coating techniques (e.g. sputtering), which are generally struggling with bad adhesion due to low particle energy. [18-27]

Implantation of ablated atoms and ions in the polymer surface is only possible during the first 100 to 500 laser pulses, when a coating thickness of 5 to 15 nm on the surface is reached. Afterwards, the dense film is too thick allowing the passing of impinging high-energetic ions (< 1000 eV), as revealed by computer simulations by the SRIM (stopping and range of ions in matter) [28] technique. [14] Thus, 20 nm film thickness was used as an ideal value for the time-consuming XPS investigations. However, already during the deposition of the first monolayers, the implantation depth is more and more decreasing. In the PLD the quick, dense covering of the surface is strongly influenced by the dense slower expanding, lower energetic (thermalized) second particle flux in the laser ablated plasma, which atoms are not implanted but deposited directly at the interface (coating surface). Afterwards, changes of the interface bonding between polymer chains and film atoms caused by high-energetic particles are negligible. [29]

5. Wrinkling and Buckling in Growing PLD Coatings on Polymers

To show the effects of these unfavourable material properties of TPU on the adhesion of PVD coatings, in a first step conventional magnetron sputtering at room-temperature was applied to deposit Al coatings. As shown in Fig. 3, the low-energetic sputtering deposition conditions (general industrially-used coating parameters) result in the formation of buckles (hemispherical caps as well as telephone-cord buckling) in the coating due to the low adhesion of the abrupt interface between the polymer surface and the coating. Besides the low adhesion, the missing of diffusion during film growth at the low sputtering temperatures prevents the filling the pores between the growing columns. Thus, the cohesion in the porous coating is diminished. Cracking mainly appears at the borders of these columns under quite low (tensile) stresses. As mentioned above, such stresses appear at the base of the buckles, where the delamination front at the interface is found. The high stress concentration and the sharp crack tip support this interfacial cracking. Combined with the porous structure, crack growth perpendicular to the surface is evident, splitting off the buckle (Fig. 5a). Higher energetic sputtering conditions (Fig. 5b) result in denser growth structures, allowing the elongation of hemispherical buckles to one direction (telephone cord buckling).

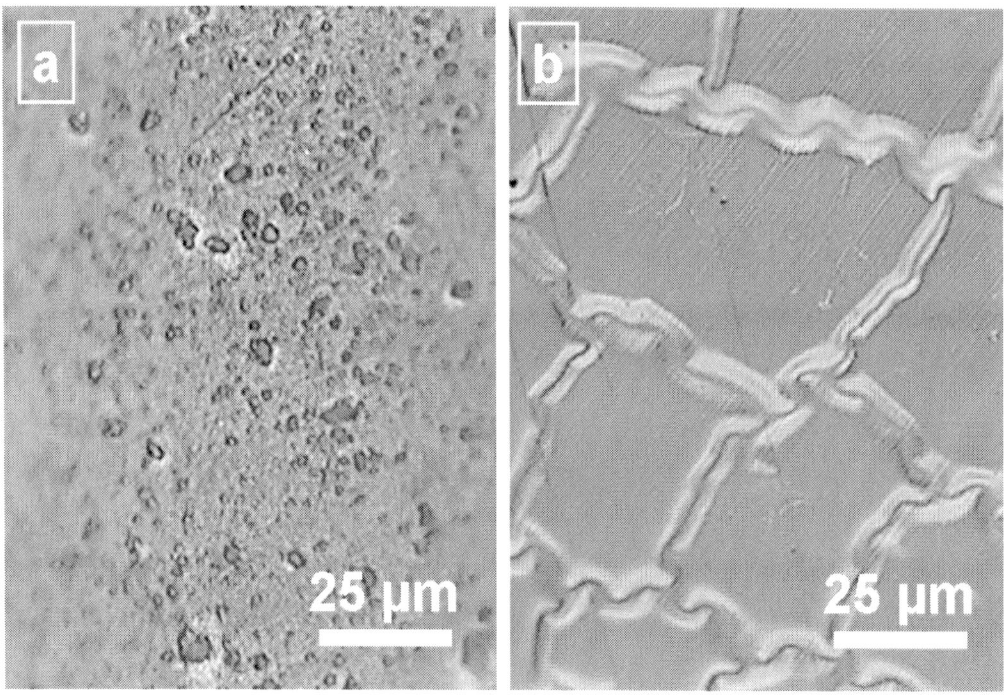

Figure 5. Light microscopy images of the TPU surface coated by room-temperature DC magnetron sputtering with 1 µm Al: (a) hemispherical buckle formation (dark gray, round areas due to splitting-off of the hemispherical caps) (very low energetic deposition conditions), (b) telephone cord buckling (slightly higher energetic deposition conditions).

Generally, the buckle formation and cracking propagate during the practical use of the coatings under the cyclic mechanical and/or thermal stresses, gradually destroying the coating and, hence, the function of the surface. To overcome this problem, the application of the room-temperature PLD technique was thought to be advantageously, which allows high adhesion of the deposited films by the implantation of the metal atoms in the polymer surface (interface) during the first steps of film growth (formation of a pseudodiffusion interface).

But compared to sputtered coatings, possessing a compressive stress level (on steel substrates) between 0.5 and 3 GPa, the room-temperature deposited PLD coatings generally have quite higher compressive residual stresses due to their high-energetic deposition conditions (particle energies up to 1000 eV) of the pulsed PLD plasma .

As expected for the higher stress-level of the PLD coatings, a visible deformation of the substrate-coating compound occurs even at a very low thickness of the coating (~ 0.2 μm) on the high-elastic TPU substrates. This deformation is clearly obvious in the light microscopically topography images of various metal film on TPU in Fig. 6. Under the coating stresses, wrinkles or buckles are formed on the whole surfaces, which appearance is quite similar than described in the theories above. [58] These topographical units of blister-like appearance seem to be fractals, which sizes are strongly dependent on the mechanical properties of the coating. Thus, a close correlation to the wrinkle formation theory, shown above (eq. 4) [42] seems to be evident: As higher the Young's modulus of the coating, as higher is the distance between the formed wrinkles (wavelength of the wrinkles). Additionally, the increase of the film thickness increases this wavelength (coarsening of the wrinkle topography), shown in Fig. 6 for the various metal coatings too. In a first approach such behaviour can only be mechanically explained by the increasing resistance to deformation of thicker and stiffer coatings.

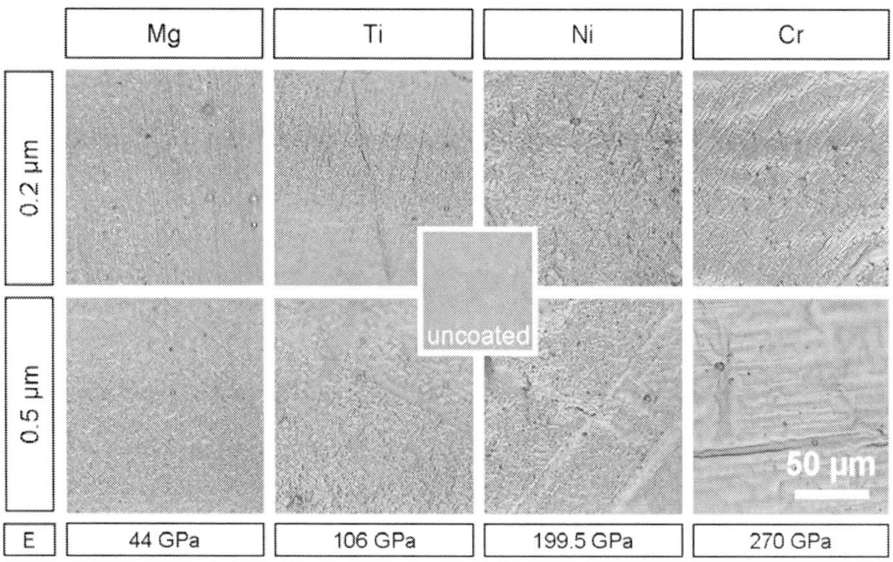

Figure 6. Light microscopy images of the uncoated and coated TPU surfaces. The industrially-scaled PLD technique at room temperature was applied to deposit metals of very different Young's moduli (Mg, Ti, Ni, Cr) on the TPU. The Young's moduli given for the metals represent the elastic behaviour of bulk metals.

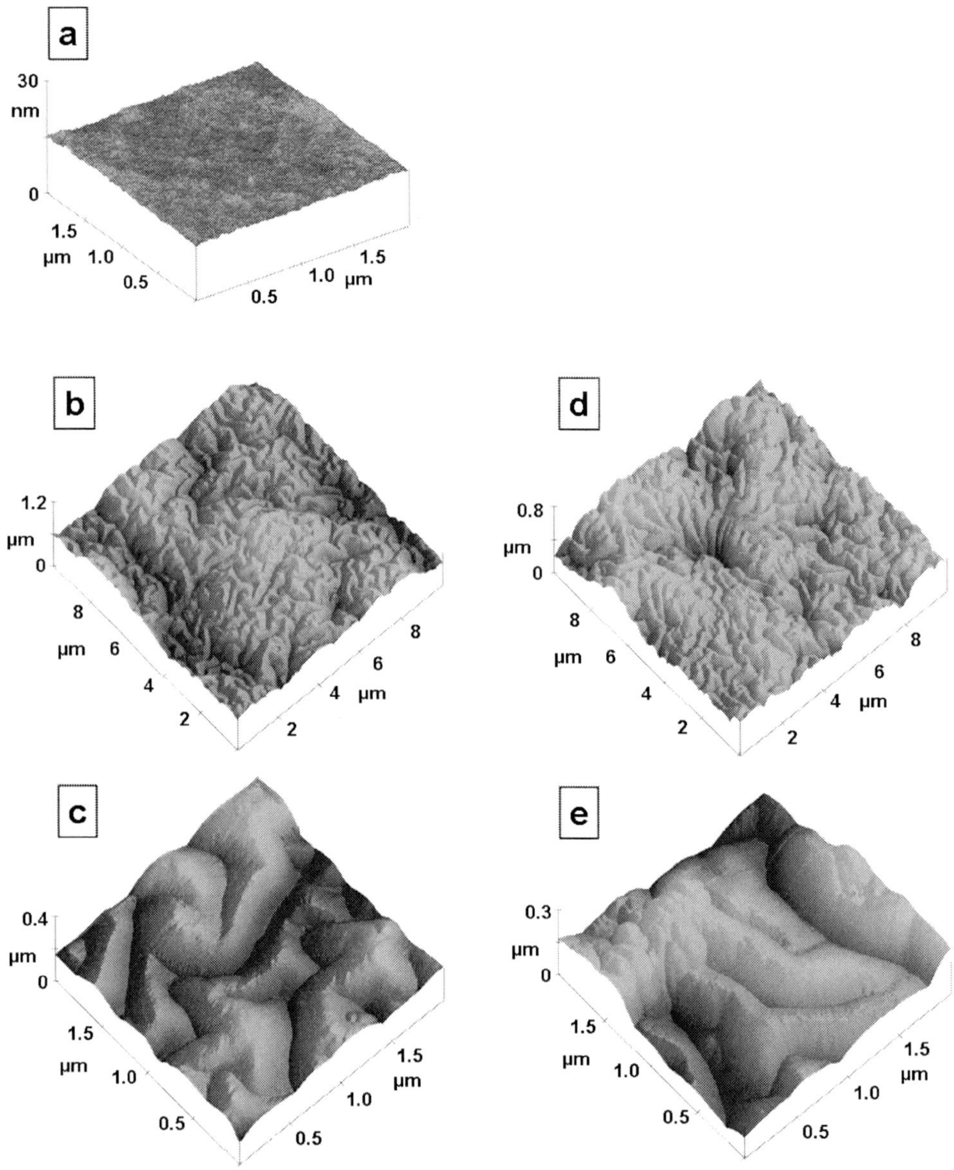

Figure 7. AFM images of (a) the uncoated TPU surface and (b-e) TiN PLD coated TPU surfaces (b,c: film thickness: 0.2 µm; d,e: 0.5 µm TiN). Fig. c and e are details taken from Fig. b and d, resp.

In the following, the description of the development of the TPU surface topography is focussed on titanium nitride (TiN) room-temperature PLD coatings. TiN was chosen as coating material, because of the higher biocompatibility of Ti compared to all of the other investigated metals. Compared to pure Ti, the addition of N increases the hardness and wear resistance (abrasive and adhesive wear), but leads to the implications of higher elastic modulus and higher residual growth stresses. Metallic titanium, deposited at room temperature by the PLD technique on ferritic steel substrates, possesses a Young's modulus of about 110 GPa, a hardness of about 10 GPa and a compressive growth stress of about –0.9 GPa. In contrast, PLD deposited TiN shows an modulus of about 250 GPa, a hardness between 32 and 35 GPa and compressive stresses of about –11.9 GPa [36] without any stress-decreasing interface layers. A detailed view on the surface topography is given for two TiN coatings of different thickness on TPU in Fig. 7. The coarsening of the wrinkle topography at higher coating thicknesses, as shown in Fig. 6, is also evident in the AFM micrographs. But additionally two very distinct sizes of wrinkles are evident: (1) A network of large wrinkles (4 to 6 μm wavelength), which are comparable to that in the light microscopy images in Fig. 6 and, additionally, (2) very small wrinkles (0.5 to 1 μm wavelength). To understand the mechanisms of their formation, all mechanisms during film growth on the soft polymer surfaces have to be taken into account. Then, a prevention of their formation seems to be possible.

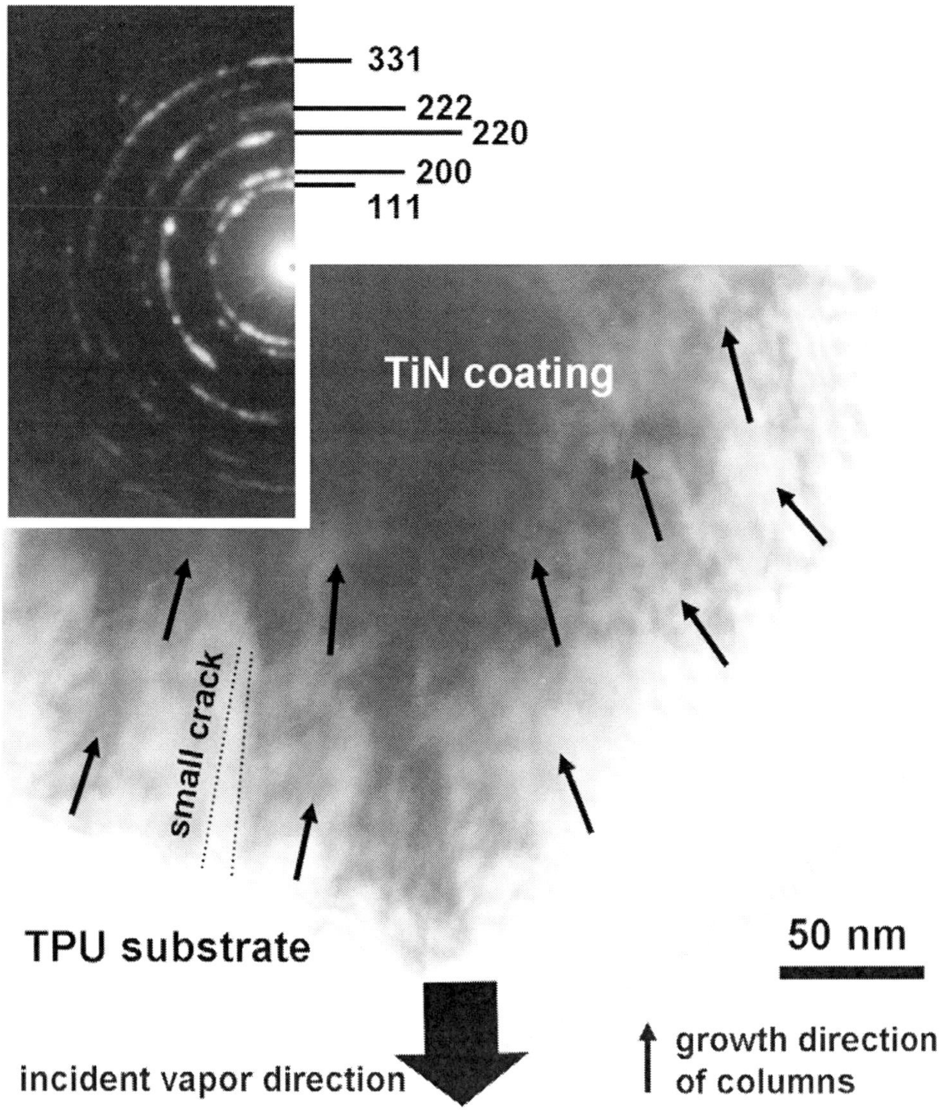

Figure 8. TEM image of the surface of the TPU substrate and the covering 0.5 μm thick TiN coating as well as of the electron micro-diffraction pattern of the TiN coating.

In all practical wrinkling and buckling problems, imperfections are expected to be important as initial nuclei of the deformation. [53, 59-66] As shown in Fig. 7a, some imperfections are evident on the surface, possibly causing small differences in the stress field of a coating. Even in the initial steps of deposition the growth stresses are evident, which continuously increase in subsequent film growth. The effects are shown in Fig. 8: Initially, the surface of the TPU substrate was flat (see Fig. 7a) and aligned perpendicular to the incident plasma flux. Hence, the growth of the micro-columns starts normal to the polymer surface (marked by the small arrows). But in the TEM image (Fig. 8) this direction is not parallel to the incident direction of the pulsed plasma. Thus, the change of the growth direction of the micro-columns can only be caused by the deformation of the substrate during growth. Such a substrate deformation is evident in the TEM image.

The initially flat surface starts to form wrinkles under the increasing growth stress. As higher the film thickness is, as higher becomes the growth stress and as higher becomes the elastic deformation of the TPU substrate surface region. Due to the similar incident direction of the pulsed plasma during the whole film growth, the orientation of the micro-columns in the microstructure follows the main particle stream during their growth, remaining in their curved main axis.

Due to the high elastic deformation of the substrate–film compound, the stress as well as the lattice strain in the TiN coating are quite low. This is evident by the low residual stress level (about -1.5 GPa in TiN on TPU compared to about –11.9 GPa on ferritic steel without any stress-decreasing interface layer) and by the formation of the (200) growth texture. This texture component possesses the lowest surface energy in TiN and predominates in low-stressed TiN films. This is evident in the electron diffraction pattern in Fig. 8, but also in the results of the XRD investigations shown in Fig. 9.

Apart from the low crystallinity (observable by the low intensity) of the TiN coatings on the amorphous PU substrate, the (200) pole figure (Fig. 9b) shows not one, but several (200) intensity maxima around its centre. In contrast to one large maxima, indicating a large amount of grains with very low difference of their lattice orientation from e.g. the (200) orientation parallel to the surface (e.g. found in TiN coating on stiff substrate materials), this behaviour is caused by only some distinct grains (columns) with a rather exact deviation of their lattices from the (200) orientation parallel to the surface. Finding a reason for this phenomenon is difficult, but the TEM images of the TiN coating in Fig. 10 can give hints. In contrast to Fig. 8, which represents the growth conditions at the interface during deformation of the substrate, Fig. 10 shows the processes, appearing during growth of thicker coatings on the polymer. Due to the increase of the growth stress at increased film thickness, the soft substrate – not able to withstand this stress – increases its deformation. Except to a possible deflection of the whole coating surface, the formation of wrinkles on the surface leads to a stress relaxation at higher coating thickness. But as thicker the film is, the more difficult is the deformation of the coating due to its decreasing elasticity. Because of the high adhesion by the pseudodiffusion interface on the TPU surface, finally through-thickness cracking of the coating (and not cracking at the interface (delamination) like for magnetron sputtered coatings) appears which is clearly shown in Fig. 10. After through-thickness cracking a total stress minimization in the substrate surface by deformation is possible. Both edges of the crack are free of stress. An increase of the stress is evident at increased distance to the crack. [67] The large deformation of the very soft substrate surface results in a wide dehiscing of the crack edges (the microcolumns on both sides of the crack). The drastically change of the direction of the grains, initially grown under normal incidence of the pulsed plasma, can cause the higher number of maxima of the (200) direction around the centre in the pole figure in Fig. 9b.

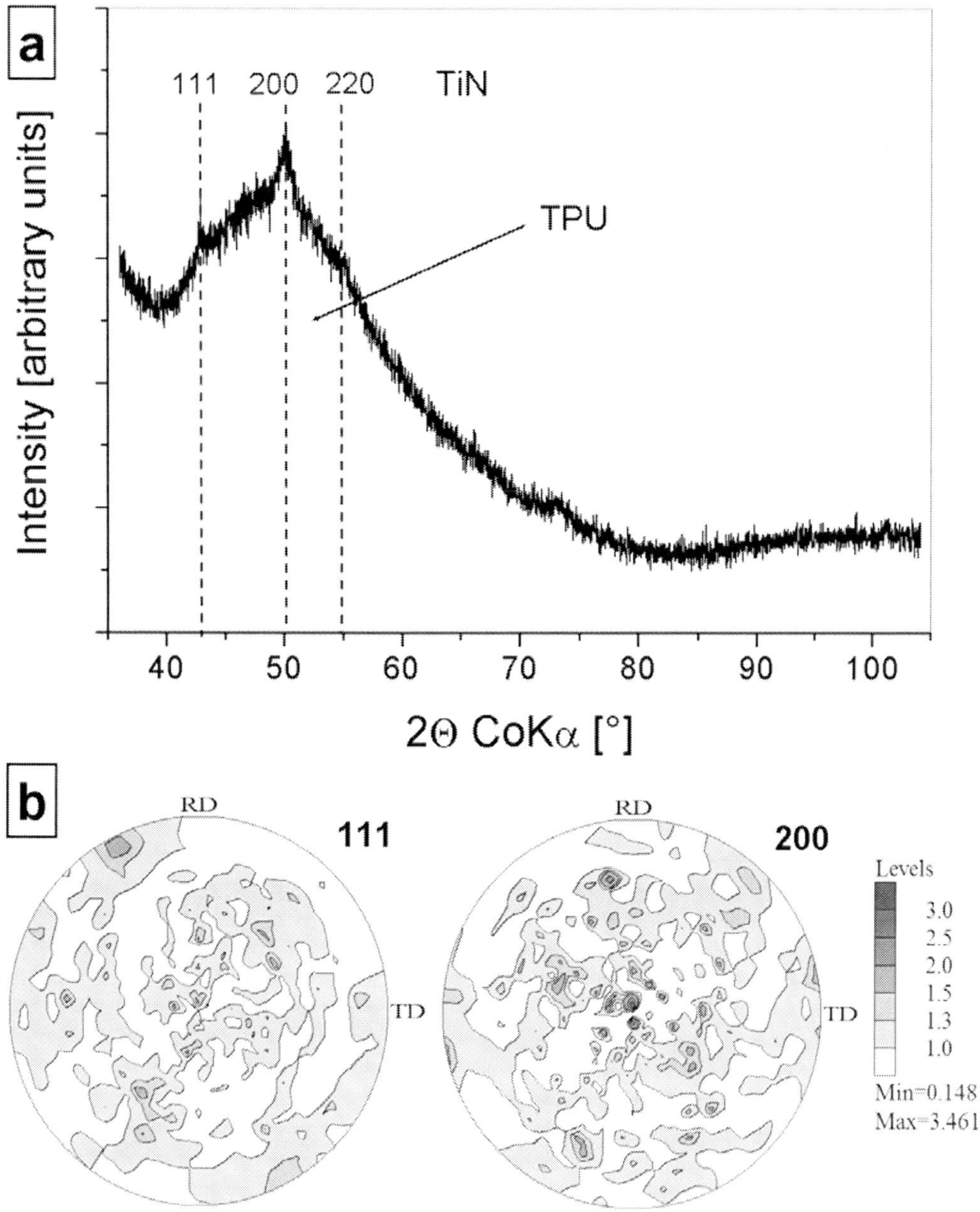

Figure 9. XRD results of a 0.5 μm thick TiN coating on a TPU substrate. (a) Phase analysis, (b) texture analyses of the (111) and (200) orientation.

This mechanism of cracking was revealed by FEM on a simplified model of a hard coating on a very soft (polymer) substrate with a large imperfection on the surface. The size of the imperfection was chosen in the same size as the size of the wrinkles on the coated surface. To simulate the growth stresses during growth, the hard coating in the model was divided in two halves (Fig. 11a): The upper halve, representing the surface-near hard coating, applied the load on the lower half, the link to the TPU substrate. The load of the upper layer was applied by its swelling. Such a swelling can be compared to an increase of the compressive stress in this (and, hence, in the adjacent) layer of the coating. By introducing the wrinkle (defect) in the model, the stress maxima and, thus, the cracking are focussed in this region. The comparison of the path of the crack in the FEM model (Fig. 11b) and in the TEM micrograph (Fig. 10) results in a very similar behaviour. The crack runs not directly from the interface of the polymer and the hard coating to the centre, but shows some inclination. Additionally, cracks appear in the FEM model along the interface and are also found in the TEM investigations of the interface-near coating region (small crack in Fig. 8).

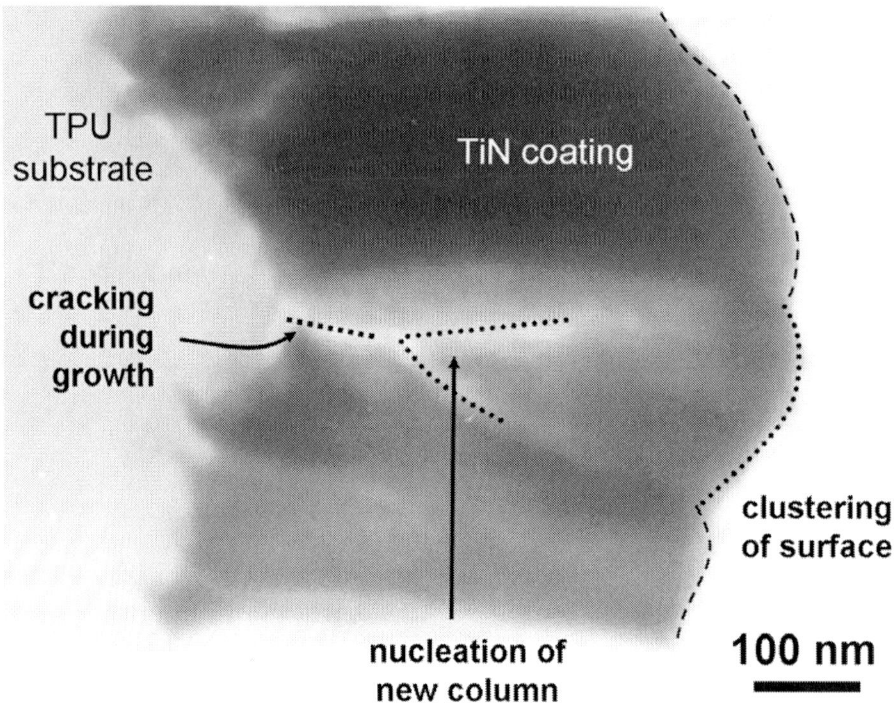

Figure 10. TEM image of a 0.5 μm thick TiN coating on a TPU substrate. Cracking during growth with subsequent nucleation of a new column is indicated.

If such a crack opens normal to the substrate surface and runs to the top of the coating, resulting in the wide dehiscing of the crack edges, it is clear, that subsequent deposition results in filling of the free volume by coating atoms. Shadowing of the main particle stream is caused by the crack edges, resulting in a lower density of the coating in the cracked regions, evidenced in the constrast differences in the TEM image (Fig. 10). Additionally, the nucleation and growth of a new micro-column starting from the crack is evident. This new column provides the cohesion of the coating after cracking due to the healing of the defect during subsequent growth. The well known growth phenomenon in such columns results in the formation of a cluster at the surface. Such cracking is found very often, apparently near each wrinkle. Thus, the very fine topographical structures (wavelength 0.5 to 1 μm), evident in the AFM images in Fig. 7, can be explained by this cracking theory.

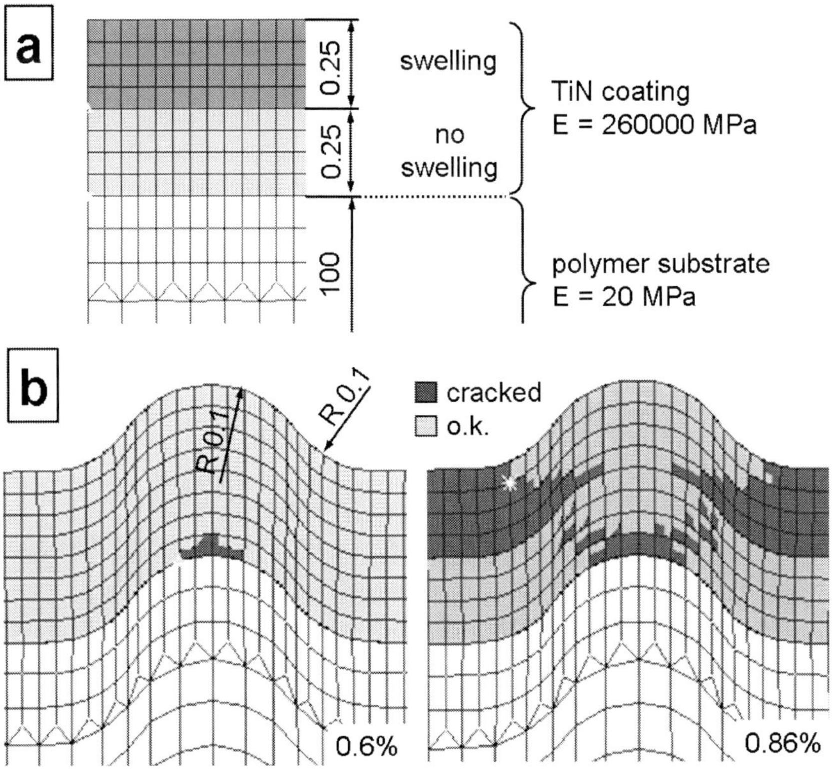

Figue 11. Finite element model (FEM) for cracking of a hard, stiff coating (ideally brittle) on a soft, deformable (polymer) substrate (ideally elastic). The materials data for the model is taken for TiN and TPU, resp.. No possibility of delamination at the interface is specified. The applied load is realized by swelling of the upper layer of the coating. (a) Description of the model and the materials data (all measures in μm). (b) Cracking of the TiN coating at 0.6 % (left image) and 0.87 % (right) swelling and a surface defect (buckle) of 300 nm curvature. The first crack appears at 0.53% swelling exactly in the centre.

The growth stress appearing in subsequent growth helps to close the initial crack normal the interface by small plastic deformation too (evident by the not parallel crack edges near the interface in the TEM micrograph, Fig. 8). Thus, the cohesion in the coatings is guaranteed, but during subsequent growth the residual stress is increasing again. The theory of Volynskii et al. [42] about wrinkling phenomena let conclude that due to the lower elasticity of the thicker coating the size of the wrinkles (meaning their wavelength) is increasing. Hence, cracking is found to occur in a lower density on the coating surface, but the cracks are opening wider. This theory allows the explanation of the formation of the second fraction, the larger wrinkles (4 to 6 μm wavelength) in the AFM images (Fig. 7). The cracking of these wrinkles leads now to weak regions (paths) in the coating, which are not fully filled with subsequently deposited atoms. Thus, the cohesion in these paths is lower than in the rest of the coating.

The high cohesion between the micro-columns in the coating, the distinct weak paths in the coating, and the high adhesion to the interface is revealed by tensile tests performed in a SEM too. According to Bazhenov et al. [67] and Volynskii et al. [68], tensile tests of coated polymer surfaces lead to a fragmentation of the coating under the tensile load. Fragmentation of the coatings is found in the direction of tension. An increase in strain leads to a decrease in fragment length by cracking into halves. Thus, the stress in the fragments is decreased. In addition, applying the tensile load leads to a contraction in the perpendicular direction – thus, compressive stresses are introduced in the compound. These compressive stresses result in a wavy structure – a sinusoidal folding of the soft substrate. The coating has to follow this deformation. Due to the folding additional tensile stresses appear in the coating. If the stress in these waves is too high, cracking can occur.

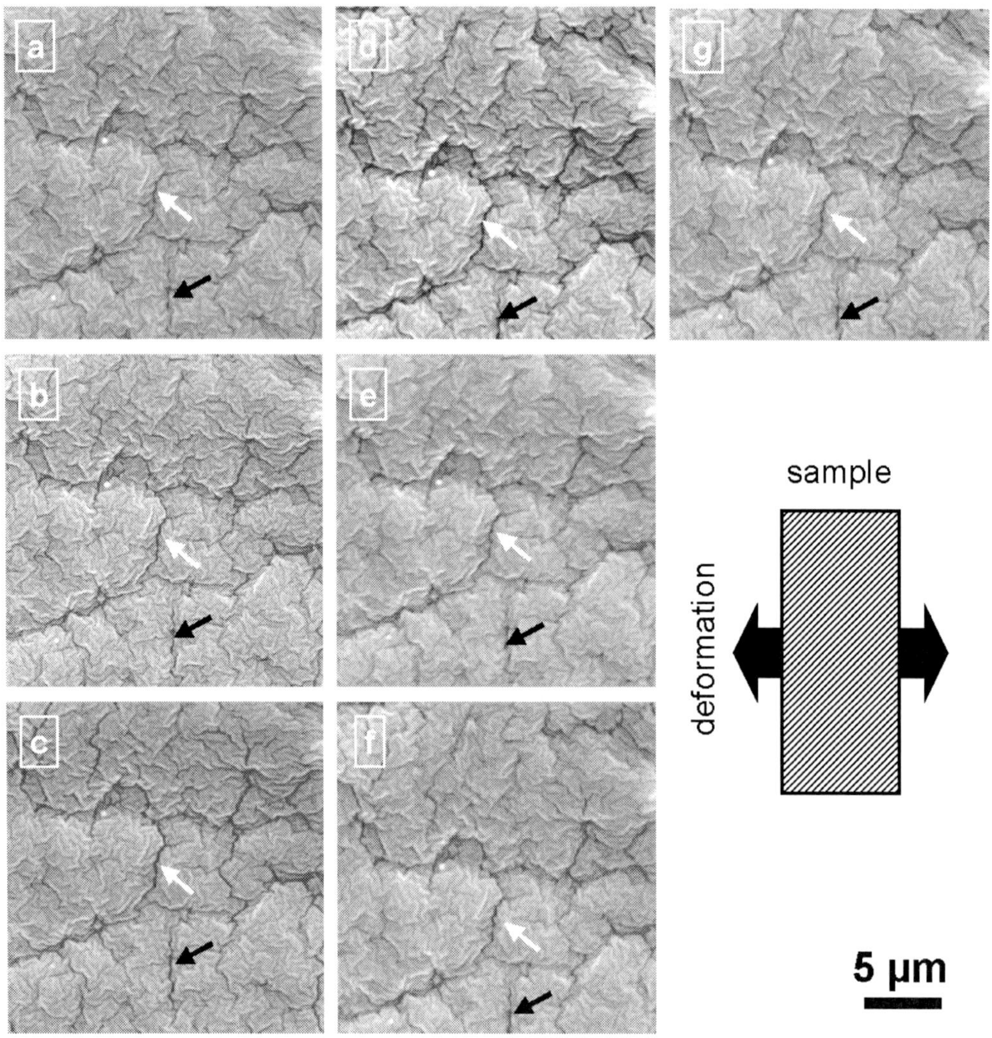

Figure 12. Deformation behaviour of the TiN coated TPU substrate (coating thickness: 0.2 µm) under tensile load in the SEM: (a) initial state (unstressed), (b) 5 % deformation, (c) 10 % deformation, (d) stress release (unstressed), (e) 5 % deformation in 2nd cycle, (f) 10 % deformation in 2nd cycle, (g) stress release after 2nd cycle (unstressed). The black and white arrows indicate the same positions in all images.

In the following, the mechanisms of TiN coating deformation at low and medium loads (elongation) on the TPU substrate are discussed. The effects of small deformations up to 10 % (related to the soft TPU substrate) on the stiff TiN coating are shown in Fig. 12: Starting from the unstressed state (Fig. 12a), the cumulative tensile force results in the opening of only some distinct cracks (Fig. 12b and c, resp.). These cracks (which run across the coating, as shown later) occur along paths, which are already evident on the unstressed surface. The chosen paths are situated quite perpendicular to the loading direction. In comparison to the AFM images (Fig. 7), such paths seems to be pre-damaged regions, e.g. by the described cracking phenomenon occurring in late steps of growth (formation of large wrinkles). In contrast, the packets of a size between 4 to 6 μm in square do not crack inside (high cohesion despite the crack network healed in subsequent growth), only at their border. Thus, the whole deformation in the surface region of the TPU substrate is concentrated near these crack paths. Due to the high elastic deformability of TPU (up to 550%), the small (but even localized) deformation seems to be nearly fully reversible, which is evident in Fig. 12d. The reversibility is supported by the nearly stress-free coating regions at the crack edges formed in the late phase of film formation (formation of the larger wrinkle network). Subsequent reloading to the same load (Fig. 12e and f) after the first release shows no differences in the cracking behaviour compared to the first cycle. The missing of delaminations during the cyclic tensile test proves the high adhesion of the PLD TiN coating (pseudodiffusion interface) on the polymer substrate too.

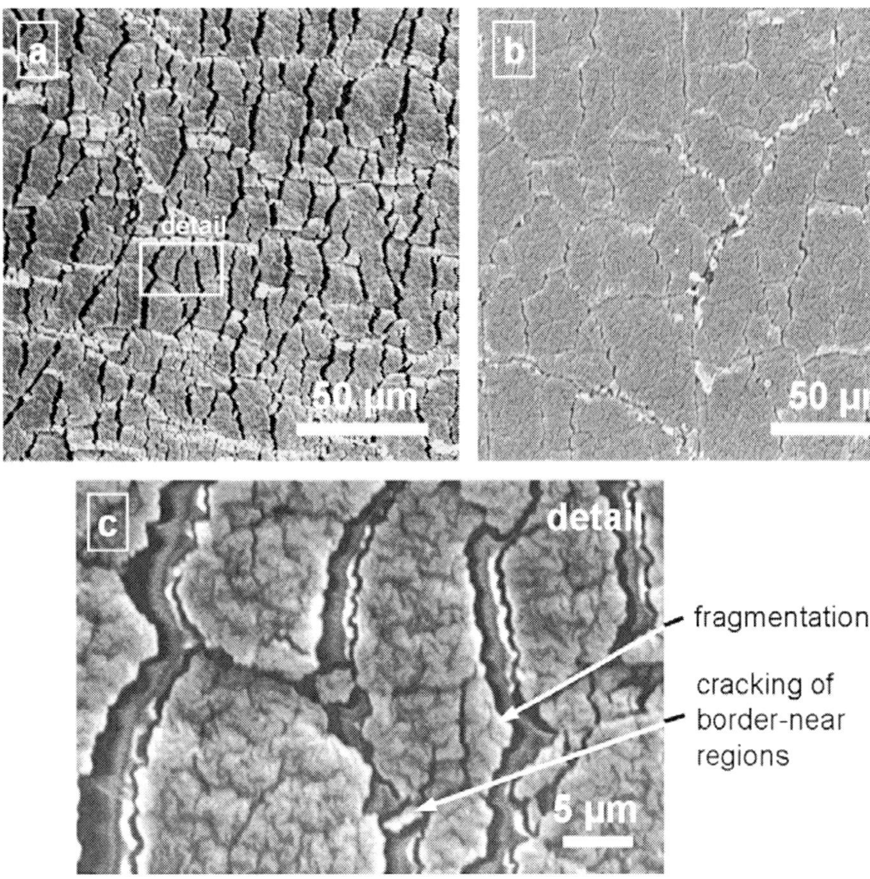

Figure 13. Deformation behaviour of the TiN coated PU substrate (coating thickness: 0.5 μm) under tensile load in the SEM: (a) 20 % deformation, (b) stress release (unstressed), (c) detail of (a).

Increasing the deformation to 20 % changes the behaviour of the compound during unloading. The higher load results again in the formation of cracks between the TiN packets of high cohesion (Fig. 13a). However, the very high bonding strength at the interface results due to high deformation in these bands on the polymer surface in the cracking at the border of the TiN packets. The propagation of large tensile strain from the beneath the borders of the well cohesive TiN packets softens these regions. The formation of compressive strain and stress in the perpendicular direction supports the formation of cracks in these regions. Additionally, the fragmentation of the coating, as described above, starts in the cohesive packets. As a result, some free-standing columns are found along these highly deformed bands. Loosing adhesion on the TPU of such a free-standing column is much easier than for a large packet – even at the high deformation in the bands – resulting in loose particles in the crack (Fig. 13c). Because of the misfit during unloading, these particles prevent total close of the crack. Thus, the coating does not cover the whole polymer surface after one loading cycle with high deformation . Fig. 13b proves that these phenomenons occur only in the high-deformed bands. A full closing of the cracks occurs in the regions of lower stressed deformation bands as explained for the lower deformation before. Thus, it's evident, that not all deformation bands are on the same strain level, possibly mainly dependent on their orientation to the tensile load direction and their distance to the next deformation bands.

CONCLUSION

Concluding these findings for low- and high-energetic deposition conditions (magnetron sputtering and PLD, resp.), in both cases unwanted changes of the coated polymer surfaces are evident. For sputtered films, delamination by buckling removes the film fully from the polymer surface. For the PLD coating, wrinkling results in rough and partly cracked, but well adherent coatings. This was revealed by findings of pseudodiffusion interfaces between PLD coatings and polymer substrates as shown for polyurethane, even at room temperature coating. Additionally, slight dissociation of the polymeric chain molecules followed by changes of the binding in the polymer surface by the implanted metal atoms were evident. To improve the properties of the coated polymer surfaces, all surface deformations have to be eliminated. This means, growth stresses in the coatings have to be diminished drastically, but simultaneously a lost in adhesion strength is unacceptable. Such techniques were developed for PLD coating and will be presented in following works.

ACKNOWLEDGMENTS

Financial support of this work by the Austrian Federal Ministry of Traffic, Innovation and Technology, the Austrian Industrial Research Promotion Fund (FFF), the Government of Styria and the European Union is highly acknowledged. Additionally the author thanks the following colleagues for common discussion and for their excellent work in film characterization and FEM modeling: B. Major, L. Major, R. Major (Institute for Metallurgy and Materials Science, Polish Academy of Science, Cracow), P. Lacki (Faculty of the Mechanical Engineering and Computer Sciences Institute of Metal Working, Quality

Engineering and Bioengineering, Czestochowa University of Technology), W. Waldhauser, R. Berghauser (Laser Center Leoben, JOANNEUM RESEARCH Forschungsgesellschaft mbH).

REFERENCES

[1] Riester, M.; Bärwulf, S.; Lugscheider, E.; Hilgers, H. *Surf. Coat. Technol.* 1999, 116-119, 1179.

[2] Liston, E. M. *Adhes.* 1989, 30, 199.

[3] Nowak, S.; Küttel, O. M. *Mater. Sci. Forum* 1993, 140-142, 705.

[4] Gröning, P.; Küttel, O. M.; Collaud-Coen, M.; Dietler, G.; Schlapbach, L. *Appl. Surf. Sci.* 1995, 89, 83.

[5] Burkstrand, J. M. *J. Appl. Phys.* 1981, 52, 4785.

[6] Atanasoska, L. J.; Anderson, S. G.; Meyer, H. M.; Lin, Z.; Weaver, J. H. *J. Vac. Sci. Technol.* 1987, A 5, 3325.

[7] Bodö, P.; Sundgren, J. *J. Vac. Sci. Technol.* 1988, A 6, 2396.

[8] Akhter, S.; Zhou, X.-L.; White, *J. M. Appl. Surf. Sci.* 1989, 37, 201.

[9] Bou, M.; Martin, J. M.; Le Mogne, *Th. Appl. Surf. Sci.* 1991, 47, 149.

[10] Lackner, J. M.; Waldhauser, W.; Fian, A.; Major, B.; Major, L.; Major, R. *Thin Solid Films* 2005, submitted.

[11] Akhsakhalyan, A. D.; Biryutin, B. A.; Gaponov, S. V.; Gudkov, A. A.; Luchin, V. I. *Sov. Phys.-Techn. Phys.* 1982, 27, 969.

[12] Akhsakhalyan, A. D.; Biryutin, B. A.; Gaponov, S. V.; Gudkov, A. A.; Luchin, V. I. *Sov. Phys.-Techn. Phys.* 1982, 27, 973.

[13] Lackner, J. M. *Innovative Coating by Pulsed Laser Deposition*; PhD thesis; University of Leoben, Austria, 2003.

[14] Lackner, J. M. Industrially-scaled hybrid Pulsed Laser Deposition at Room Temperature; *OREKOP* s.c., Krakow, Poland, 2005.

[15] Mattox, D. M. ASM *Handbook Vol. 5 – Surface Engineering*; ASM – International Society for Metals, Materials Park, Ohio, 1994, 538.

[16] Lackner, J. M.; Waldhauser, W.; Berghauser, R.; Ebner, R.; Kothleitner, G. *Surf. Coat. Technol.* (2005) submitted.

[17] Zaporjtchenko, V.; Strunskus, T.; Behnke, K.; Bechtoldsheim, C. v.; Thran, A.; Faupel, F. *Microelectr. Eng.* 2000, 50, 465.

[18] Lee, J. H.; Cho, J.-S.; Koh, S.-K.; Kim, D. *Thin Solid Films* 2004, 449, 147.

[19] Park, S. K.; Han, J. I.; Kim, W. K.; Kwak, M. G. *Thin Solid Films* 2001, 397, 49.

[20] Kukla, R.; Hofmann, D.; Ickes, G.; Patz, U. *Surf. Coat. Technol.* 1995, 74-75, 648.

[21] Schiller, S.; Kirchhoff, V.; Schiller, N.; Morgner, H. *Surf. Coat. Technol.* 2000, 125, 354.

[22] Bertrand, P.; Lambert, P.; Travaly, Y. *Nucl. Instr. Meth. Phys. Res.* 1997, B 131, 71.

[23] Dietzel, Y.; Przyborwski, W.; Nocke, G.; Offermann, P.; Hollstein, F.; Meinhardt, J. *Surf. Coat. Technol.* 2000, 135, 75.

[24] Friedrich, J. F.; Unger, W. E. S.; Lippitz, A.; Koprinarov, I.; Kühn, G.; Weidner, St.; Vogel, L. *Surf. Coat. Technol.* 1999, 116-119, 772.

[25] Guzman, L.; Miotello, A.; Voltolini, E.; Adami, M. *Thin Solid Films* 2000, 377-378, 760.
[26] Kupfer, H.; Wolf, G. K. *Nucl. Instr. Meth. Phys. Res.* 2000, B 166-167, 722.
[27] Popovici, D.; Klemberg-Saphieha, J. E.; Czeremuszkin, G.; Sacher, E.; Meunier, M.; Martinu, L. *Microelectronic Eng.* 1997, 33, 217.
[28] Ziegler, J. F.; Biersack, J. P. *SRIM – The Stopping and Range of Ions in Matter;* Software-Version 2003.20, 2003.
[29] Lackner, J. M.; Waldhauser, W. Jahrbuch *Oberflächentechnik* 2005, 128,
[30] Knotek, O.; Löffler, F.; Wolkers, L. *Surf. Coat. Technol.* 1994, 68/69, 176.
[31] Rickerby, D. S.; Eckold, G.; Scott, K. T.; Buckley-Golder, I. M. *Thin Solid Films* 1987, 154, 125.
[32] Thornton, J. A. *J. Vac. Sci. Technol.* 1974, 11, 666.
[33] Weissmantel, C.; Schürer, C. *Thin Solid Films* 1979, 61, 5.
[34] Pan, A.; Green, J. E. *Thin Solid Films* 1981, 78, 25.
[35] Lackner, J. M.; Waldhauser, W.; Ebner, R.; Major, B.; Schöberl, T. *Thin Solid Films* 2004, 453-454, 195.
[36] Lackner, J. M. *J. Phys.* 2006, accepted.
[37] Gong, X.-Y.; Clarke, D. R. *Oxidation Metals* 1998, 50, 355.
[38] Biot, M. A. *J. Appl. Mechanics* 1937, 4, A1.
[39] Biot, M. A. *Mechanics of Incremental Deformation*, Wiley, New York, NY, 1965.
[40] Biot, M. A. *Proc. Roy. Soc. Lond., Ser.* 1957, A 242, 1231.
[41] Landau, L. D.; Lifshitz, E. M. *Theory of Elasticity*, Pergamon Press, Oxford, Great Britain, 1986.
[42] Stoney, G. G. *Proc. R. Soc. Lond.*, 1909, Ser. A 82, 172.
[43] Eymery, J. P.; Boubeker, B. *Material Lett.* 1994, 19, 137.
[44] Eymery, J. P.; Denanot, M. F. *Surf. Coat. Technol.* 1996, 80, 251.
[45] Kinbara, A.; Baba, S.; Matuda, N.; Takamisawa, K. *Thin Solid Films* 1981, 84, 205.
[46] Thouless, M. D. *J. Am. Ceram. Soc.* 1993, 76, 2936.
[47] Gioia, G.; Ortiz, M. *Adv. Appl. Mech.* 1997, 33, 119.
[48] Yu, H. Y.; Kim, C.; Sandy, S. C. *Thin Solid Films* 1991, 196, 229.
[49] Gille, G.; Rau, B. *Thin Solid Films* 1984, 120, 109.
[50] Matuda, N.; Baba, S.; Kinbara, A. *Thin Solid Films* 1981, 81, 301.
[51] Hutchinson, J. W.; Thouless, M. D.; Liniger, E. G. *Acta Met. Mat.* 1992, 40, 295.
[52] Jain, S. C.; Harker, A. H.; Cowley, R. A. *Philos. Mag.* 1997, 75, 1461.
[53] Pundt, A.; Nikitin, E.; Pekarski, P.; Kirchheim, R. *Acta. Mat.* 2004, 52, 1579.
[54] Mylvaganam, K.; Zhang, L. C. *Thin Solid Films* 2003, 425, 145.
[55] Iyer, S. B.; Harshavardhan, K. S.; Kumar, V. *Thin Solid Films* 1995, 256, 94.
[56] Moon, M.-W.; Chung, J. W.; Lee, K.-R.; Oh, K. H.; Wang, R.; Evans, A. G. *Acta Mater.* 2002, 50, 1219.
[57] Tolpygo, V. K.; Clarke, D. R. *Mater. Sci. Eng.* 2000, A 278, 151.
[58] He, M. Y.; Evans, A. G.; Hutchinson, J. W. *Mater. Sci. Eng.* 1998, A 245, 168.
[59] Mumm, D. R.; Evans, A. G. *Acta Mater.* 2000, 48, 1815.
[60] Evans, A. G.; He, M. Y.; Hutchinson, J. W. *Acta Mater.* 1997, 45, 3543.
[61] Evans, A. G.; Hutchinson, J. W.; He, M. Y. *Acta Mater.* 1999, 47, 1513.
[62] Hutchinson, J. W.; He, M. Y.; Evans, A. G. *J. Mech. Phys. Solids* 2000, 48, 709.

[63] Ito, Y. M.; Rosenblatt, M.; Cheng, L.Y.; Lange, F. F.; Evans, A. G. In. *J. Fract.* 1981, 17, 483.

[64] Bazhenov, S. L.; Valynskii, A. L.; Alexandrov, V. M.; Bakeev, N. F. *J. Polymer Sci.* 2002, 40, 10.

[65] Volynskii, A. L.; Bazhenov, S.; Lebedeva, O. V.; Ozerin, A. N.; Bakeev, N. F. *J. Appl. Polymer Sci.* 1999, 72, 1267.

In: Adhesive Properties in Nanomaterials, Composites ... ISBN: 978-1-61209-268-3
Editors: K. A. Wilkinson and D. A. Ordonez

Chapter 4

ADHESIVE BONDING OF HYDRO-THERMALLY MODIFIED WOOD

Andreja Kutnar*
University of Primorska,
Primorska Institute for Natural Sciences and Technology,
Muzejski trg 2, 6000 Koper, Slovenia; ILTRA d.o.o.,
Celovska cesta 268, 1000 Ljubljana, Slovenia

ABSTRACT

In the past years considerable increase in the hydrothermal modification of wood was observed. Mostly the heat treatments are performed to change the hygroscopicity of wood. Furthermore, densification processes are utilizing the hydrothermal treatments. A key factor in the efficient utilization of timber resources is the adhesive bonding of wood, since manufacturing of wood based composites depends on forming bonds between individual wooden elements. Wood-based composites offer several advantages over sawn wood, such as the utilization of waste material, better distribution of non-homogeneities, and control of the product properties in the manufacturing process. Therefore the efficient utilization of hydro-thermally modified wood depends on its adhesive potential. The combined effects of temperature and moisture modify the properties of the polymeric components of wood and its porous structure. Wood tissue is exposed to high temperatures that can cause surface inactivation. Hydrothermal treatment could reduce the surface free energy and thus result in the poorer wettability of the modified wood surface. Furthermore, penetration and spreading of the resin could be influenced by hydrothermal treatment. In spite of numerous studies of hydro-thermally modified wood, the adhesion potential of hydro-thermal treated wood has not been studied extensively in the past. The aim of this chapter is to provide literature review of aspects like surface properties of hydro-thermally modified wood related to bondability, wetting, and penetration. Finally, future directions regarding efficient application of hydro-thermally modified wood including densified wood in polymer composites are discussed.

* E-mail address: andreja.kutnar@upr.si

1. Introduction

Wood is a valuable renewable resource, used in numerous applications. However, it has some less desirable properties, such as vulnerability against insect and fungi attack, dimensional instability, relatively low resistance against weathering, etc. There are many ways to overcome the negative characteristics of wood. One possibility is to produce new wood-like materials using wood just as a raw material. For instance, there are numerous reports on the production, characteristics and applications of chemically or thermally treated wood (e.g. Epmeier et al. 2004; Rowell 2006). In general, modified wood exhibits improved dimensional stability and increased resistance against wood pests. The poor mechanical properties of low-density wood can be modified and improved by wood densification processes. Various methods of wood densification have been reported (Seborg et al. 1945; Inoue et al. 1993; Dwianto et al. 1999; Navi and Girardet 2000; Blomberg and Persson 2004). Kamke and Sizemore (2008) developed a method for wood densification using a viscoelastic thermal compression (VTC) process. All new wood-like materials are used in the same way as normal, unprocessed wood. This means that they should allow surface finishing with commercial wood coatings as well as bonding with common wood adhesives. However, surface characteristics (surface free energy, surface morphology) of wood-based materials could be different from those of untreated wood and might affect bonding and surface finishing.

The use of various adhesives as auxiliary has enabled a enormous diversity of applications. However, the adhesive bonding continues to be one of the most challenging area of the research in wood science owing to new wood-like materials the high variable and complex nature of raw material. Wood exhibits enormous variability due to the differences in its composition. Porous structure, surface roughness, chemical heterogeneity of wood, and hydroscopic nature are just some of the main surface properties that are strongly associated with the adhesive penetration.

2. Hydro-Thermally Modified Wood

The hydro-thermally modified wood with the most changed structure is densified wood. The concept of wood densification has been known for at least a century, when the first patented procedures of densification appeared (Kollman et al. 1975). Wood can be densified by impregnating its void volume with polymers, molten natural resins, waxes, sulphur, and even molten metals (Kultikova 1999). Wood can also be densified by compressing it in a transverse direction under conditions that do not cause damage to the cell wall (Kollman et al. 1975). The obtained densified wood products have increased strength, stiffness and hardness.

Densification by the transverse compression of wood consists of four stages: softening, compression, setting, and fixation of the compressed state. The viscoelastic nature of wood plays an important role in compression and densification. When the wood temperature is above the glass transition temperature (T_g) of its amorphous polymers, then large deformations can occur without fractures. The springback effect, which is one of the main problems associated with the densification process, can be eliminated by steaming or heating, which can induce permanent fixation of the compressive deformation (Kutnar and Kamke 2010).

Wood densification alters wood properties. In addition to increasing density, hydrothermal modification results in reduced equilibrium moisture content (Kollmann and Cote 1968; Metsä-Kortelainen et al. 2006), improved dimensional stability (Hillis 1984; Hsu et al. 1988; Inoue et al. 1993; Esteves et al. 2007), and improved biological durability (Boonstra et al. 2007). However, Skyba et al. (2008) found that THM densification treatment increased the resistance of spruce wood but not of beech wood to degradation by soft-rot fungi. Furthermore, Kutnar et al. (2010) determined that VTC densification did not change decay resistance to *Pleurotus ostreatus* and *Trametes versicolor*. Additionally, the effect of heat and moisture in the densification process generates chemical changes in the wood structure, which can be seen as significant color changes (Tjeerdsma et al. 1998; Koch et al. 2003; Sundqvist et al. 2006; Varga and van der Zee 2008). Moreover, the morphology of densified wood can change considerably, depending on the degree of densification (Blomberg et al. 2006; Kutnar et al. 2009). The results of microscopic examination of VTC wood have established that the VTC process adequately takes into account the plasticization of wood (Kutnar et al. 2009). Thus the desired degree of compression is achieved without destroying the micro-cellular structure of the wood (Figure 1). Deformations in the VTC wood are largely the result of the viscous buckling of cell walls without any fracture taking place. The strength and stiffness of the wood material are therefore increased approximately in proportion to the increase in density (Kutnar et al. 2008a). The density profile of the VTC wood varies with the degree of densification (Kutnar et al. 2009). Different density profiles are the consequence of temperature and moisture gradients, and their relationship to the glass transition of the wood cell wall at the time the compression stress is applied. Density distribution is also visible on the cross-section of VTC specimens. Lower and higher density layers can be distinguished by the different quantities of void areas across the thickness.

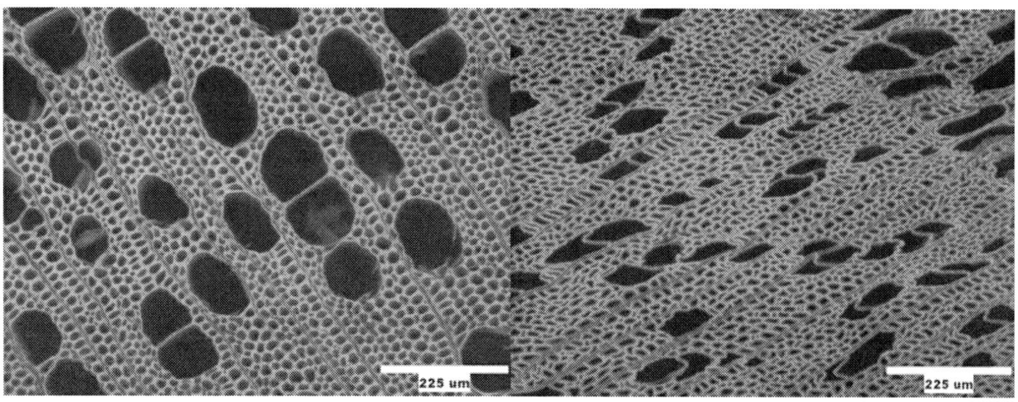

Figure 1. Photomicrographs of the control wood (left) and the VTC wood with a 132% degree of densification (right); the sections were obtained from water-soaked specimens. Compression was applied in the vertical direction as shown here.

The simultaneous effects of heat, moisture and compression generate chemical and physical changes in the wood structure, which have a strong influence on the surface properties of densified wood (Jennings et al. 2006; Kutnar et al. 2008b). Kutnar et al. (2008b) studied the changes in surface chemistry of the VTC wood by means of Fourier transform infrared spectroscopy (FTIR). Since the chemical changes of the VTC wood were not observed in the FTIR spectra, the research compared the VTC densification process to thermal modification, consisting of a moist step and a dry step. The polymerization reactions of the lignin were defined as the major chemical changes induced by the VTC process. In addition, the wettability study on the basis of the contact angle of water, using the Wilhelmy plate method, showed that the VTC wood surface exhibits hydrophobic behavior (Table 1) (Kutnar et al. 2008b).

Table 1. Average (n=10) advancing contact angle obtained with the different test liquids (the standard deviation is shown in parentheses) (Kutnar et al. 2008b; Petrič et al. 2009).

	Contact angle (°)			
Sample	Water	Formamide	Diiodomethane	PF adhesive
Control	71.6 (7.6)	38.7 (19.6)	5.4 (9.2)	133.7 (7.8)
VTC 63%	90.9 (5.7)	79.2 (6.6)	48.5 (6.5)	126.3 (5.6)
VTC 98%	93.6 (10.3)	75.4 (3.7)	60.9 (7.0)	127.8 (3.0)
VTC 132%	93.1 (6.2)	81.8 (5.0)	58.0 (5.17)	126.3 (3.0)

3. Adhesive Bonding of Densified Wood

A key factor in the efficient utilization of timber resources is the adhesive bonding of wood. Adhesive bonds are formed in a wood structure by a combination of the spreading of the adhesive along the surface, and by penetration, wetting, and solidification. For two wood adherents to be held together with maximum strength, the adhesive must diffuse over and into each surface in order to make contact with the molecular structure of wood, so that the intermolecular forces of attraction between the adhesive and the wood can become effective. Adhesive penetration can then be defined as the spatial distance from the interface of the adjoining substrate (Sernek et al. 1999). The porous and highly permeable structure of wood together with the pit network that connects adjacent cells are causing the variations in adhesive penetration. The pattern of adhesive penetration influences the performance of adhesive bonds in wood.

Using today's technologies, almost any material can be bonded. The results of research have shown that densified wood can be bonded using a polymeric diphenylmethane diisocyanate adhesive, a B-stage phenol-formaldehyde film adhesive (Jennings et al. 2005) or by liquid phenol-formaldehyde (PF) adhesives (Kutnar et al. 2008a; Kutnar et al. 2008c). However, increased density, and consequently decreased porosity, of densified wood has an effect on the flow and penetration of the adhesive. The void volume of densified wood can be drastically decreased, depending on the degree of densification. The depth and direction that an adhesive flows in densified wood is therefore much different than in virgin wood. The exposure of wood to high temperatures during densification processing alters the surface properties of wood (Kutnar et al. 2008b). The wettability of wood, which has been exposed to high temperature or heat treated, is decreased (Pétrissans et al. 2003; Sernek et al. 2004; Follrich et al. 2006; Gérardin et al. 2007). The wood becomes hydrophobic, less polar and rather repellent to water, which can prevent waterborne adhesive systems from being able to adequately wet the surface. Surface free energy determined according to the Owens Wendt Rabel and Kaelble (OWRK) theory decreases significantly due to the VTC densification process, while the level of densification had a limited influence on the surface free energy (Table 2) (Kutnar et al. 2008b).

Table 2. The surface free energy of the control and VTC wood obtained by the OWRK approach (Kutnar et al. 2008b).

	Surface free energy [mJ/m^2]		
Sample	γ_s	$\gamma_{sv}{}^d$	$\gamma_{sv}{}^p$
Control	53.4	50.6	2.7
VTC 63%	35.5	35.1	0.3
VTC 98%	28.6	28.1	0.5
VTC 132%	29.8	29.7	0.0

It appears that the hydrophobic character of the VTC wood samples is a consequence of the chemical changes in the wood due to hydrothermal conditioning during the densification process and the related reduction in the surface free energy of densified wood, as suggested by Jennings et al. (2006). In addition, it has to be taken into account that the contact angle can be also influenced by the wood surface morphology (Gardner 1996; Mantanis and Young 1997; Bico et al. 2002; Packham 2003). The control wood surface is rougher and has greater porosity then the VTC wood, which could result in lower contact angle of the test liquids on the control wood. Furthermore, the VTC process significantly changes the morphology of the VTC wood (Kutnar et al. 2009). Densification to a higher degree is achieved by a larger reduction in the void spaces of the wood. Therefore, it was expected that the contact angles of the test liquids would depend on the degree of densification. However, the results revealed that the degree of densification had minor and limited effect on the contact angle of the VTC wood (Table 1).

Indeed, the surface free energy of VTC densified poplar wood decreased from the original value of 53 mJ/m^2 to 28-36 mJ/m^2 due to the densification process, as calculated by the Owens-Wendt-Rabel-Kaelble (OWRK) approach (Table 2). The total surface free energy of the control wood was relatively close to the data in the literature (de Meijer et al. 2000). The decrease in the surface free energy on the VTC wood was the consequence of increased hydrophopic character of heat-treated wood due to hemicellulose's degradation and reorganization of the lignocellulosic polymeric components of wood (Gérardin et al. 2007). Apparently, a larger reduction in the void spaces of the wood due to higher degree of densification did not affect the surface free energy. Both the dispersion and the polar components of the surface free energy were higher in the case of the control wood than the VTC wood. Previous studies on the surface free energy of heat treated wood have, however, reported a slight increase in the dispersion component after heat treatment (Gérardin et al. 2007; Jennings et al. 2006). The decrease of the dispersion component of the surface free energy of the VTC wood is most likely the result of surface ageing (Kutnar et al. 2008b).

Since the wettability study with the Wilhelmy plate method showed that the VTC wood surface exhibited hydrophobic behavior, the concern rose on potential increase of contact angle of certain adhesive types. However, contact angles of the liquid PF adhesive in dependence of the degree of densification of VTC poplar wood did not indicate statistically significant differences in the contact angle among the VTC wood specimens with different degrees of densification (Table 1). However, the contact angle on the control wood was significantly higher than that obtained on the VTC 63% wood and the VTC 132% wood, respectively. The results could be the consequence of the contact angle measurements of the high viscous PF adhesive, where it is difficult to assess the equilibrium. The air trapped in pores could account for high values of contact angles with PF on the control wood. Since the pores of the VTC wood are reduced, the air trapped appearance was less pronounced in the contact angle measurements on the VTC wood surface (Petrič et al. 2009). Contrary to the expectation, based on the increase of water contact angles on VTC wood, a slight decrease of contact angles of the PF adhesive was observed. This is a promising and interesting result, opening possibilities of using the adhesive, just as in the case of normal, unprocessed wood.

3.1. Adhesive Bonding of the VTC Wood

Increased density, and consequently decreased porosity, of the VTC wood affects adhesive flow and penetration. Studies of bonding properties of the VTC wood established that wood densification using the VTC process did not affect the ability of the surface of the VTC wood to properly bond with PF adhesives (Kutnar et al. 2008a; Kutnar et al. 2008c). The hydrophobic behavior of the VTC wood surface and the decreased effective penetration (EP) of the PF adhesive into the VTC wood were not crucial for this adhesion (Figure 2 and Figure 3).

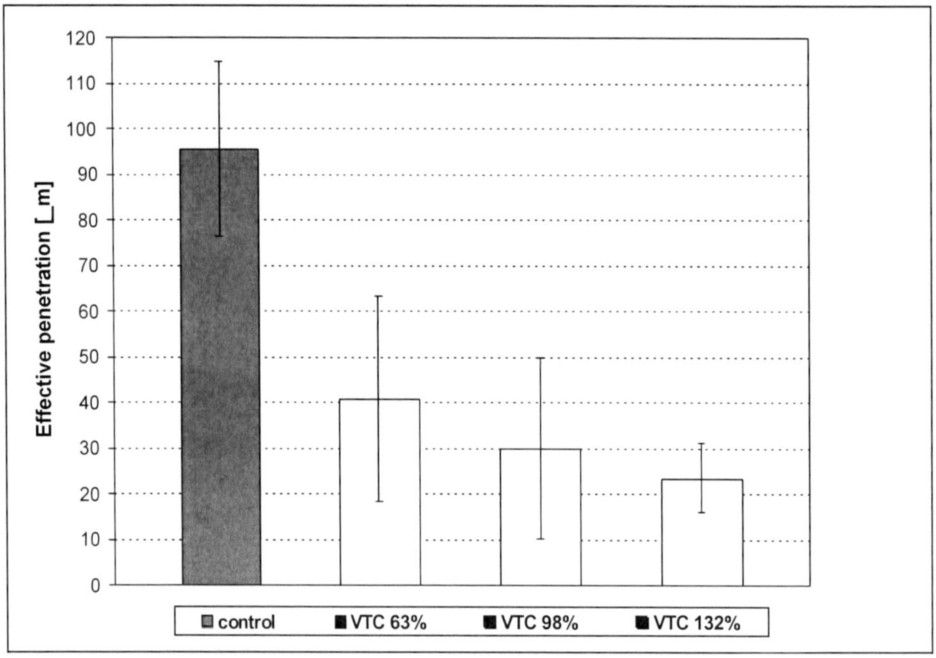

Figure 2. The effective penetration of PF adhesive into the control and VTC wood specimens of different degrees of densification (Kutnar et al. 2008c).

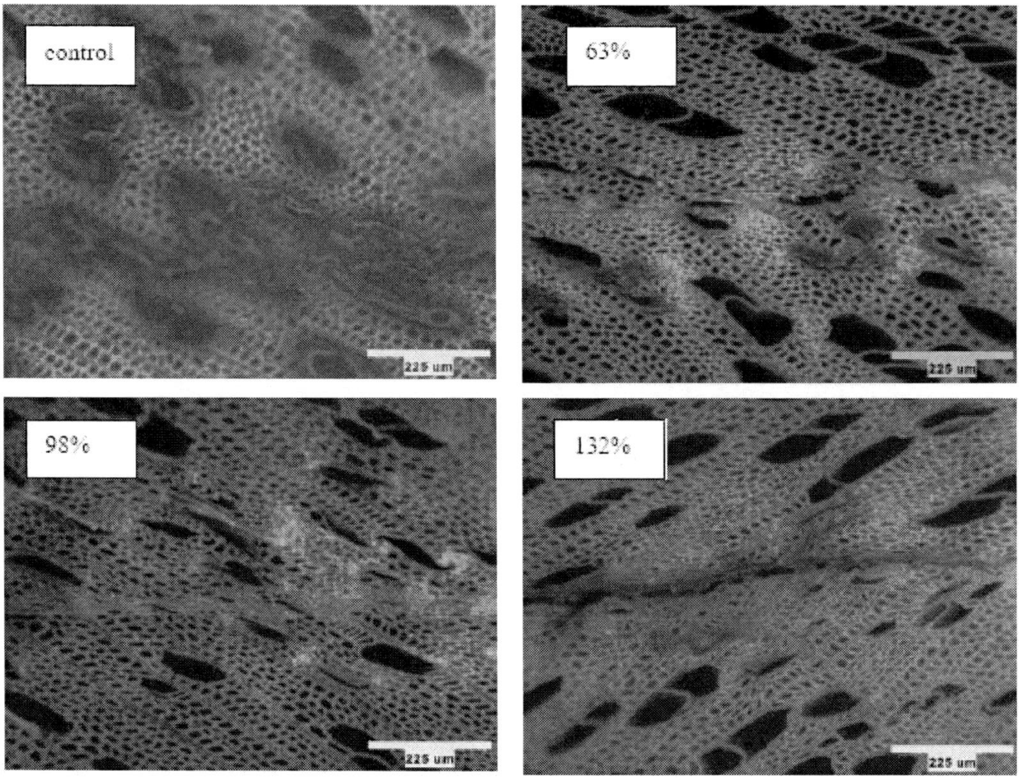

Figure 3. The PF adhesive penetration in the control and the VTC wood specimens of different degrees of densification (Kutnar et al. 2008c).

Effective bonding of the VTC wood was achieved. The PF adhesive bond strength of the densified VTC specimens was similar or better than that of the control (undensified) specimens.

Although the EP depended on the level of densification, the bond performance between specimens having different degrees of densification could not be classified.

This could be the consequence of the formation of formaldehyde and lignin polymerization reactions during the VTC densification process due to the hydrothermal treatment used (Kutnar et al 2008b). Additionally, Kutnar et al. (2008a) determined that bonding of the VTC wood to the untreated control wood is not problematic, although penetration of the adhesive is mainly into the control wood. Due to closed micro-cellular structure of the VTC wood, the majority of the PF adhesive penetrated into the cell lumens of the control wood (Figure 4).

The adhesive followed the path of least resistance. Hence the EP of the PF adhesive differs significantly among the control wood and VTC wood test specimens. Surface related phenomena such as surface inactivation, which can lead to poor bonding (Sernek et al. 2004), is not an issue in the VTC process.

The VTC process is accomplished in a pressurized system; so there is little potential for the migration of extractives to the surface (Kamke 2006).

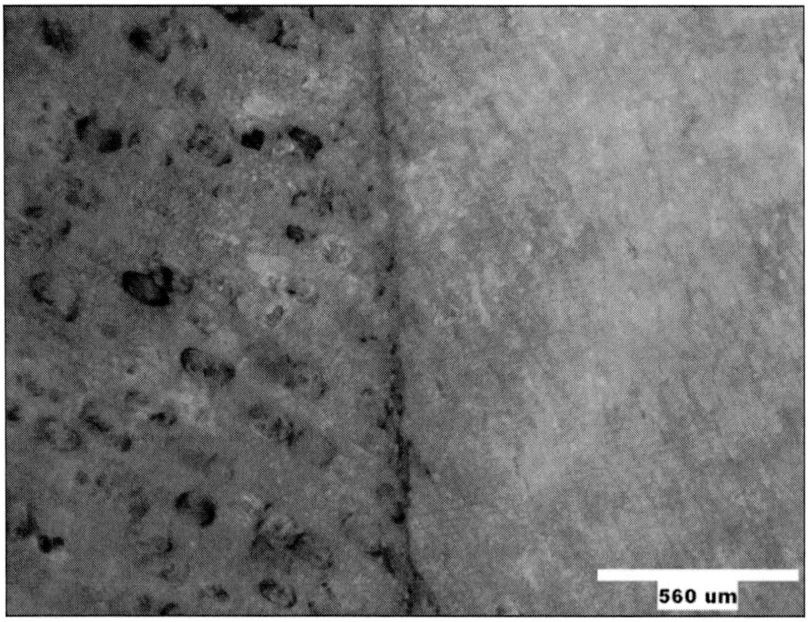

Figure 4. Cross-sectional view of the bonded untreated wood and of VTC wood with a 132% degree of densification. The adhesive is clearly distinguishable from the wood; the dark regions indicate the presence of the PF adhesive. The specimen was embedded with polymerized linseed oil, and polished using abrasive diamond disks (Kutnar et al. 2008a).

4. Application of Densified Wood in Wood Composites

The purpose of wood composites is the transformation of materials into items of greater utility and value. The ideal wood-based composite is either stronger, lighter, more durable, attractive, and/or cheaper than competing materials. The basic element for composite wood products can be timber (glulam), veneers, veneer strands (plywood, LVL), flakes, strands (OSB, OSL), wood particles (particleboards), fibre (paper, fibreboards), adhesives, and other additives, and finishing materials. Wood-based composites offer several advantages over sawn wood, such as the utilization of waste material, better distribution of non-homogeneities, and control of the product properties in the manufacturing process. Additionally, adhesives can increase the strength and stiffness of the composite, since they can effectively transfer and distribute stresses.

The development of the VTC process has made it possible to develop a new wood-based composite with low-density undensified wood in the core and high density VTC wood in the faces (Kamke and Sizemore 2008; Kutnar et al. 2008a). The high density VTC wood surface layers resist in-plane and bending loads, whereas the low density core determines the product thickness, moment of inertia, and carries the shear loads. Kutnar et al. (2008a) manufactured four different 3-layer laminated composites (Figure 5 and 6). The 3-layer control composites (0-0-0) were manufactured from undensified wood. The outer laminas were 2.5 mm thick, whereas the core was 6 mm thick. The 3-layer VTC composites (63-0-63, 98-0-98, 132-0-132) were manufactured from densified and undensified wood. The VTC wood laminas of different degree of densification (63, 98, and 132%) were placed in the two outer layers, and a 6 mm thick piece of undensified wood was placed in the core layer (Figure 5 and 6).

Figure 5. The 3-layer VTC composite (98-0-98).

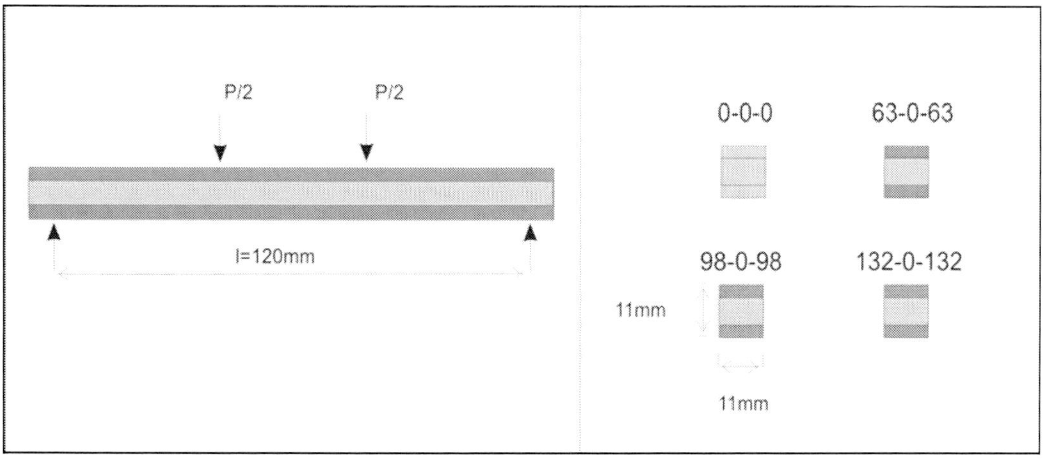

Figure 6. The geometry of the 3-layer VTC composite specimens. Left: The specimens for the 4-point bending tests of the 3-layer composites; Right: The composition (transverse view) of the four different 3-layer composites (0 – the control wood; 63 – the VTC wood with a 63% degree of densification; 98 - the VTC wood with a 98% degree of densification; 132 - the VTC wood with a 132% degree of densification) (Kutnar et al. 2008a).

The four-point bending tests (Figure 6) of four different 3-layer composites made from undensified wood and VTC wood showed that the modulus of rupture (MOR) and the modulus of elasticity (MOE) of the 3-layer VTC composites were significantly improved due to the increased density of the VTC wood in the face layers of the 3-layer composites (Kutnar et al. 2008a). Kamke (2006) tested three-layer VTC composites with VTC wood in the face layers and a piece of untreated wood, of the same species and original thickness, in the core layer. The obtained MOE was compared with the MOE values for undensified specimens. The increase in MOE for the three-layer composite was 2 to 3 times greater than the corresponding increase in mean density of the composite.

Conclusions

The presented results confirmed that low-density wood species can be successfully used for the production of structural wood-based composites, if they are densified by means of the VTC process. VTC wood is therefore definitely a product that could be easily adopted by and have a significant impact on the wood products industry. Structural wood-based composites may be produced and still have yield strength and stiffness properties that are comparable to any structural wood-based product that is currently produced (Kamke 2006). Furthermore, low-density and juvenile wood with poorer mechanical properties can be used in new high-performance wood-based composite products for structural components in buildings, transportation systems, and casework. However, a key factor in the efficient utilization of timber resources is the adhesive bonding of wood. Adhesives can increase the strength and stiffness of the composite by effectively transferring and distributing stresses. The ideal wood-based composite is stronger, lighter, more durable, attractive, and/or cheaper than competing materials. Wood modification treatments significantly change the structure and composition of wood, therefore adhesively bonding of wood continues to be one of the most challenging areas of research in wood science, due to the highly variable and complex nature of wood as raw material. The most important features of wood strongly associated with adhesion and bonding performance are its porous structure, the roughness of its surface, the chemical composition, and its hydroscopic nature. The pattern of adhesive penetration directly affects the performance of adhesive bonds in wood. Therefore, first step towards efficient application of hydro-thermally modified wood, especially of densified wood, in polymer composites is the extensive study of structural, chemical and mechanical properties of this new wood-like material.

References

Bico, J., Thiele U., Quéré, D. (2002). Wetting of textured surfaces. *Colloids and Surfaces A: Physicochemical and Engineering Aspects*, 206: 41–46.

Blomberg, J., Persson, B. (2004). Plastic deformation in small clear pieces of Scots pine (*Pinus sylvestris*) during densification with the CaLignum process. *Journal of Wood Science*, 50,4: 307–314.

Blomberg, J., Persson, B., Bexell, U. (2006). Effects of semi-isostatic densification on anatomy and cell-shape recovery on soaking. *Holzforschung,* 60: 322–331.

Boonstra, M.J., van Acker, J., Kegel, E., Stevens, M. (2007). Optimization of a two-stage heat treatment process: durability aspects. *Wood Science and Technology*, 41: 31–57.

de Meijer, M., Haemers, S., Cobben, W., Militz, H. (2000). Surface energy determination of wood: Comparison of methods and wood species, *Langmuir,* 16: 9352–9359.

Dwianto, W., Morooka, T., Norimoto, M., Kitajima, T. (1999). Stress Relaxation of Sugi (Cryptomeria japonica D. Don) *Wood in Radial Compression under High Temperature Steam. Holzforschung*, 53(5): 541-546.

Epmeier, H., Westin, M., Rapp, A.O. (2004) Differently Modified Wood: Comparison of some selected properties. *Scandinavian Journal of Forest Research*, 19: 31–37.

Esteves, B., Marques, A., Domingos, I., Pereira, H. (2007). Influence of steam heating on the properties of pine (*Pinus pinaster*) and eucalypt (*Eucalyptus globulus*) wood. *Wood Science and Technology*, 41(3): 193−207.

Follrich, J., Müller, U., Gindl, W. (2006). Effects of thermal modification on the adhesion between spruce wood (*Picea abies Karst.*) and a thermoplastic polymer. *Holz als Roh- und Werkstoff*, 64,5: 373–376.

Gardner, D.J. (1996). Application of the Lifshitz-van der Waals acid-base approach to determine wood surface tension components, *Wood and Fiber Science*, 28(4):422–428.

Gérardin, P., Petrič, M., Pétrissans, M., Lambert, J., Ehrhrardt, J.J. (2007). Evolution of wood surface free energy after heat treatment. *Polymer Degradation and Stability*, 92(4): 653−657.

Hillis, W.E. (1984). High temperature and chemical effects on wood stability. Part 1: general considerations. *Wood Science and Technology*, 18,4: 281−293.

Hsu, W.E., Schwald, W., Schwald, J., Shields, J.A. (1988). Chemical and physical changes required for producing dimensionally stable wood-based composites. Part I: steam pretreatment. *Wood Science and Technology*, 22,3: 281−289.

Inoue, M., Norimoto, M., Tanahashi, M., Rowell, R.M. (1993). Steam or heat fixation of compressed wood. *Wood and Fiber Science*, 25(3): 224−235.

Jennings, J.D., Zink-Sharp, A., Frazier, C.E., Kamke, C.E. (2006). Properties of compression densified wood. Part 2: Surface energy. *Journal of Adhesion Science and Technology* 20(4): 335–344.

Jennings, J.E., Zink-Sharp, A., Kamke, F.A., Frazier, C.E. (2005). Properties of compression densified wood. Part 1: Bond performance. *Journal of Adhesion Science and Technology* 19(13-14):1249–1261.

Kamke, F.A. (2006). Densified radiata pine for structural composites. Maderas. *Ciencia y technologia*, 8(2): 83−92.

Kamke, F.A., Sizemore, H. (2008). Viscoelastic thermal compression of wood. *U.S. Patent* No. 7,404,422.

Koch, G., Puls, J., Bauch, J. (2003). Topochemical characterisation of phenolic extractives in discoloured beechwood (*Fagus sylvatica* L.). *Holzforschung*, 57,4: 339–345.

Kollmann, F.P., Côte, W.A. (1968). Principles of Wood Science and Technology. I. Solid Wood. Springer-Verlag Berlin. Heidelberg.

Kollman, F.P., Kuenzi, E.W., Stamm, A.J. (1975). Principles of Wood Science and Technology. Vol. II: Wood Based Materials. Springer-Verlag, New York, Heidelberg, Berlin: 139−149.

Kultikova, E.V. (1999). Structure and Properties Relationships of Densified Wood. Master Thesis. Virginia Tech, Blacksburg, Virginia, 136 pp.

Kutnar, A., Kamke, F.A., Sernek, M. (2008a). The mechanical properties of densified VTC wood relevant for structural composites. *Holz als Roh- und Werkstoff* 66,6: 439–446.

Kutnar, A., Kamke, F.A., Petrič, M., Sernek, M. (2008b). The influence of viscoelastic thermal compression on the chemistry and surface energetics of wood. *Colloids and Surfaces A: Physicochemical and Engineering Aspects* 329: 82–86.

Kutnar, A., Kamke, F.A., Nairn, J.A., Sernek, M. (2008c). Mode II fracture behavior of bonded viscoelastic thermal compressed wood. *Wood and Fiber Science* 40,3: 362–373.

Kutnar, A., Kamke, F.A., Sernek, M. (2009). Density profile and morphology of viscoelastic thermal compressed wood. *Wood Science and Technology* 43,1: 57–68.

Kutnar A., Humar M., Kamke F.A., Sernek M. (2010) Fungal decay of viscoelastic thermal compressed (VTC) wood. *European Journal of Wood and Wood Products*, DOI 10.1007/s00107-010-0432-z

Kutnar, A., Kamke F.A. (2010). Compression of wood under saturated steam, superheated steam, and transient conditions at 150°C, 160°C, and 170°C. *Wood Science and Technology*, DOI: 10.1007/s00226-010-0380-0

Mantanis, G.I., Young, R.A. (1997). Wetting of wood. *Wood Science and Technology*, 31: 339–353.

Metsä-Kortelainen, S., Antikainen, T., Viitaniemi, P. (2006). The water absorption of sapwood and heartwood of Scots pine and Norway spruce heat-treated at 170°C, 190°C, 210°C and 230°C. *Holz als Roh- und Werkstoff*, 64(3): 192−197.

Navi, P., Girardet, F. (2000). Effects of thermo-hydro-mechanical treatment on the structure and properties of wood. *Holzforschung* 54(3): 287−293.

Packham, D.E. (2003). Surface energy, surface topography and adhesion. *International Journal of Adhesion and Adhesives*, 23(6): 437–448.

Petrič, M., Kutnar, A., Kričej, B., Pavlič, M., Kamke, F.A., Šernek, M. (2009). Surface free energy of viscoelastic thermal compressed wood. Contact Angle, Wettability and Adhesion, Koninklijke Brill NV, Leiden, Vol. 6: 301–309.

Pétrissans, M., Gérardin, P., Elbakali, I., Serraj, M. (2003). Wettability of heat-treated wood. *Holzforschung*, 57(3): 301−307.

Rowell, R.M. (2006). Chemical Modification of wood: A Short Review. *Wood Material Science and Engineering*, 1: 29–33.

Seborg, R.M., Millet, M.A., Stamm, A.J. (1945). Heat-stabilized compressed wood. Staypak. Mechanical Engineering, 67: 25−31.

Sernek, M., Resnik, J., Kamke, F.A. (1999). Penetration of liquid urea-formaldehyde adhesive into beech wood. *Wood and Fiber Science*, 31(1): 41−48.

Sernek, M., Kamke, F.A., Glasser, W.G. (2004). Comparative analysis of inactivated wood surface. *Holzforschung*, 58(1): 22−31.

Skyba, O., Niemz, P., Schwarze, F.W.M.R. (2008). Degradation of thermo-hygro-mechanically (THM)-densified wood by soft-rot fungi. *Holzforschung* 62(3): 277–283.

Sundqvist, B., Karlsson, O., Westermark, U. (2006). Determination of formic-acid and acetic acid concentrations formed during hydrothermal treatment of birch wood and its relation to coluor, strength and hardness. *Wood Science and Technology*, 40(7): 549−561.

Tjeerdsma, B.F., Boonastra, M., Pizzi, A., Tekely, P., Militz, H. (1998). Characterization of thermally modified wood: molecular reasons for wood performance improvement. *Holz als Roh- und Werkstoff*, 56(3): 149−153.

Varga, D., Van der Zee, M.E. (2008). Influence of steaming on selected wood properties of four hardwood species. *Holz als Roh- und Werkstoff*, 66(1):11–18.

In: Adhesive Properties in Nanomaterials, Composites ... ISBN: 978-1-61209-268-3
Editors: K. A. Wilkinson and D. A. Ordonez

Chapter 5

THE USE OF ADHESIVE FILMS IN TRANSDERMAL AND MUCOADHESIVE DOSAGE FORMS

Kalliopi Dodou
Sunderland Pharmacy School,
University of Sunderland, UK

ABSTRACT

Thin polymeric films with adhesive ability are useful for transdermal and mucoadhesive drug delivery systems. Polymer materials with adhesive ability in their dry state are integral to the formulation of patch systems for topical and transdermal drug delivery. Such polymers are often called "pressure sensitive adhesives" due to their capacity to attach to the skin surface with the application of light pressure. In the drug-in-adhesive design the drug is mixed with the adhesive polymer to produce a thin medicated film. The adhesive performance of these films can be monitored directly using tack and peel tests and indirectly by correlation with rheological parameters. Polymers with adhesive ability following absorption of moisture are useful in the formulation of mucoadhesive films for transmucosal (*e.g.* buccal, nasal, ocular) drug delivery. Such polymers are hydrophilic (hydrogels). Following hydration, polymer chains relax and interact with mucus glucoprotein chains, primarily by hydrogen bonding. This chapter will describe the properties of the adhesive polymers used in the design of transdermal and mucoadhesive films, the mechanism of adhesion and the tests that can be applied to monitor the adhesive performance.

1. INTRODUCTION

Polymers with adhesive properties are integral to the design and function of patches and mucoadhesive dosage forms. Before engaging in the detailed description of these polymeric materials, it is important to define a few terms that will be used throughout the chapter.

Dosage forms that are applied onto the skin surface deliver the drug either topically or transdermally. Transdermal drug delivery is the delivery of an active pharmaceutical ingredient (API) into the systemic circulation via the skin. It is differentiated from topical drug delivery where the API remains within the skin layers and does not reach the systemic circulation at therapeutically effective concentrations. Topical dosage forms therefore are effective in the treatment of skin disorders, e.g. eczema, psoriasis, skin infections. Transdermal dosage forms are by contrast effective in the treatment of systemic disease conditions, and represent an alternative route to the oral and intravenous routes of drug administration. Transdermal and topical dosage forms are also called dosage forms for external application. The various types of transdermal / topical dosage forms include creams, emulsions, ointments, gels, lotions and patches. From all these dosage forms, patches are unique to this route of administration. Patches have the ability to stick to the skin surface because of the incorporation of an adhesive polymeric layer in the patch design. The first patches were developed in the 1970's for the management of motion sickness (scopolamine). Since then, extensive research in the area of transdermal drug delivery has been undertaken. Many more drugs have been subsequently formulated as transdermal patches. Examples of drugs that are present in commercial patches include: oestrogen for the management of Hormone Replacement Therapy (HRT), nitroglycerin for the treatment of angina and buprenorphine for severe pain relief.

The adhesive is the vital component of a patch system because it enables the attachment of the patch onto the skin and consequently affects the whole drug delivery process. The role and properties of such adhesive materials are the focus of this chapter and will be discussed in more detail. Other essential components of a transdermal patch are the backing layer and the release liner. The backing layer is the outer layer of the patch, exposed to the environment when the patch is applied onto the skin (Scheme 1). The backing layer protects the components of the patch from the external environment during patch application. The choice of backing layer type depends on the area and duration of application. An occlusive backing layer does not allow skin moisture loss, leading to skin hydration. This would be acceptable for small patches with a short application time (up to one day maximum). For larger patches or patches with a longer time of application (up to a week), a backing layer which allows water vapour transmission is more appropriate, because over-hydration of the skin surface may affect the adhesive ability of the patch and/or allow microbial growth. A backing layer with multi-directional stretch is required for larger patches, so as to conform to body movements. The release liner is the layer of the patch that is removed prior to application onto the skin. It protects the patch components during storage and is occlusive in order to prevent loss of volatile patch components. It is important that all patch ingredients are compatible with each other, along with the drug and any excipients.

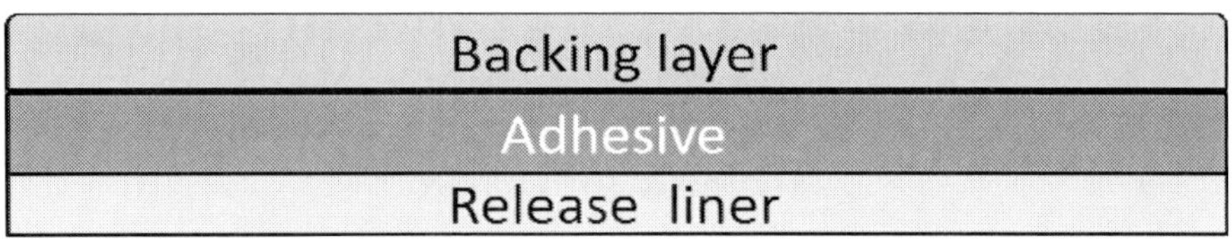

Scheme 1. Layout of the main patch components.

There are several different designs of transdermal patches. These are the matrix, the reservoir and the drug-in-adhesive designs. In the *matrix* design a gel matrix containing the drug in a dispersed or dissolved state is incorporated between the backing layer and the release liner, surrounded by a circular layer of adhesive. In the *reservoir* patch design, the drug-containing gel matrix is incorporated between the backing layer and the adhesive separated by a membrane. The role of this membrane is to control the rate of drug release from the gel matrix. In the *drug-in-adhesive* design the drug (and any other excipients e.g. penetration enhancers) are directly incorporated in the adhesive layer during the manufacture of the patch (Scheme 2). This patch design contains the least components, has the potential to be ultrathin and is therefore regarded as the state-of-the-art patch design.

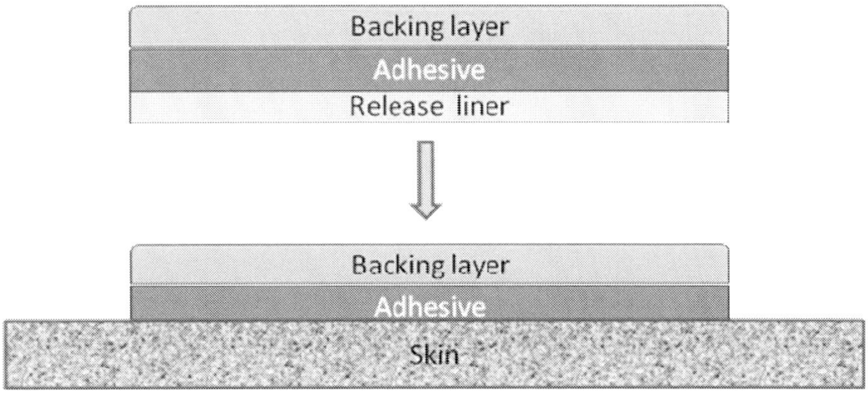

Scheme 2. Drug-in-adhesive patch design.

Mucoadhesive dosage forms belong to the wider drug delivery technique called bioadhesion. Bioadhesion is defined as the attachment or association of a drug carrier system to a biological surface. A drug carrier system is the dosage form which contains a bioadhesive polymer or ligand with the biological surface being any *internal* body surface that is accessible by oral administration e.g. stomach, intestinal lining, or by manual application e.g. buccal, nasal, ocular, rectal and vaginal surfaces. Internal body surfaces consist of epithelial cells that are covered by a mucous gel layer. There are two types of bioadhesion: mucoadhesion is the non-specific attachment of the dosage form on the mucous layer and is achieved by the incorporation of mucoadhesive polymers in the dosage form; cytoadhesion is the specific interaction between the dosage form and certain moieties of the epithelial membrane cells beneath the mucous and is achieved by the association of recognition ligands on the dosage form surface. Cytoadhesion is out of the scope of this chapter and therefore will not be described further.

It is therefore important to differentiate between adhesive films for skin application (external application) from mucoadhesive films that are intended for application to internal body surfaces.

2. Pressure Sensitive Adhesives for Transdermal Patches

The most commonly used adhesive polymeric materials in the manufacture of transdermal patches are the pressure sensitive adhesives (PSAs). Their main advantage is that they do not require a secondary method to confer adhesivity, such as water or heat and as their name implies, they attach to the skin via application of light pressure. Commonly used types of PSAs for transdermal systems are the silicones, acrylics, polyisobutylenes and triblock copolymers. For a detailed review of the different types of pressure sensitive adhesives see Tan and Pfister (1999), Satas (1999a), Venkatraman and Gale (1998), and Benedek and Feldstein (2008).

Adhesive performance is vital for therapeutic effectiveness, quality and safety of the patch. Failure of the transdermal patch to remain onto the skin surface throughout the treatment period will lead to therapeutic failure of the transdermal system. In recent years there have been frequent reports to the Food and Drug Administration (FDA) of transdermal patch systems lacking appropriate adhesivity leading to accidental detachment from the skin surface and loss of therapeutic effect (Wokovich et al., 2006). Adhesive failure can occur in three ways; a) detachment of the adhesive layer from the skin, b) detachment of the drug matrix from the adhesive layer, and c) loss of cohesive strength of the adhesive, where the adhesive splits, half remaining attached on the drug matrix or backing layer and the other half remaining attached on the skin (Wokovich et al., 2006). It is therefore necessary to understand the adhesive nature of PSAs and to be able to monitor their adhesive properties.

2.1. Mechanism of Adhesion and Adhesive Performance Monitoring

Pressure sensitive adhesives are viscoelastic materials i.e. they behave both like viscous liquids and elastic solids. Tack or bonding is a term that describes the attachment of the adhesive on the skin. According to definition, an adhesive is pressure sensitive only if tacky at room temperature; under application of light pressure the adhesive behaves like a viscous liquid and spreads onto the skin surface and once the applied pressure is removed, the adhesive remains fixed onto the skin (Satas, 1999b). Peel or debonding describes the removal of the adhesive from the skin surface. Shear strength describes the cohesive strength of the adhesive; its ability to remain attached on the skin surface without overspreading *e.g.* out of the backing layer borders, and to be completely removed from the release liner and skin surface without leaving any residues. Tack, peel and shear tests have been commonly used for monitoring adhesive performance. Tack tests measure instantaneous adhesion of a probe on the adhesive surface; peel tests measure the force required to peel the adhesive from a surface such as stainless steel or glass; shear tests measure the time taken for the adhesive to fail in a cohesive manner. For a detailed description and comparison of the various versions of each of these testing methods see Muny (1999). In addition, texture analysis has been effectively used for the evaluation of adhesive properties of rubber, acrylic and silicone adhesives; this technique can provide a unique "finger-print" graphical representation for each adhesive that is also sensitive to the addition of excipients (Ulman et al., 2006)

Monitoring of the viscoelastic nature of adhesives has been recognised as a more insightful method of determining adhesive performance (Chu 1991; Sweet and Ulman, 1997; Rohn, 1999; Satas, 1999b; Ho and Dodou, 2007). Tack, peel and shear strength can be correlated with measured viscoelastic parameters of the adhesive, using Dynamic Mechanical Analysis and Rheometry. Measured viscoelastic parameters using these techniques are G', G'', δ, J_c and η^*. The elastic or storage modulus, G', is a measure of the elastic, solid-like behaviour of the adhesive where energy is stored upon deformation. The viscous or loss modulus, G'', is a measure of the viscous, liquid-like behaviour of the adhesive, where energy is lost upon deformation. The phase angle, δ, is a measure of the adhesive's viscous and elastic contribution to its overall viscoelastic behaviour; $\delta = 0°$ corresponds to purely elastic behaviour and $\delta = 90°$ corresponds to purely viscous behaviour. Often $\tan\delta$, which is the ratio of G'' / G', is used instead of δ. Creep compliance, J_c, is defined as strain, γ, over stress, σ, (γ / σ) and is proportional to the flow of the adhesive under the application of constant stress. Complex viscosity, η^*, is a measure of the adhesive's resistance to flow.

Tack or bonding is a low rate process where the adhesive behaves like a viscous liquid in order to spread on the skin surface. Peel or debonding is a high rate process where the adhesive behaves like an elastic solid. Therefore the requirements for adhesive performance are low elastic modulus, G', at low oscillatory frequencies and high elastic modulus at high oscillatory frequencies. An oscillatory test, such as frequency sweep test at constant temperature or over a range of temperatures followed by time-temperature superposition, can be used for the measurement of G' at low and high frequencies.

Chu and Dahlquist have quantified these requirements using natural rubber blends with resin at 25° C (Chu, 1991; Dahlquist, 1999):

Chu's criteria

G' ($\omega = 0.1$ rad/s) $\sim 2 - 4 \times 10^4$ Pa

$5 < [G'(\omega = 100 \text{ rad/s}) / G'(\omega = 0.1 \text{ rad/s})] < 300$

Dahlquist's criterion of tack

$G' < 10^5$ Pa

According to Rohn (1999), tack requires low frequencies ranging from $\omega = 0.005 - 0.05$ rad/s, whereas peel requires high frequencies ranging from $\omega = 100 - 1000$ rad/s. Chu's criteria were developed using $\omega = 0.1$ rad/s as tack frequency, and $\omega = 100$ rad/s as peel frequency, allowing an easy and quick evaluation of the adhesive performance of PSAs using conventional rheometers.

The increase of elastic modulus with increasing frequency confers adhesive ability to PSAs, and is described by the elastic modulus ratio between high and low frequencies. Ratios within the defined range indicate a good balance of tack and peel properties.

Shear strength is associated with creep compliance J_c which is a measure of the flow ("cold flow") of the adhesive under the application of constant low stress. A creep test can be used for the measurement of J_c. According to Dahlquist, $J_c > 10^{-5}$ Pa^{-1} (Dahlquist, 1999) at low stress values, indicating that the adhesive should have low cold flow so as to prevent overspreading during application, but still sufficient flow that would allow spreading, also called wetting, of the adhesive during bonding. In addition, η^* can be used as an indicator of "cold flow".

Chu's and Dahlquist's criteria were deduced using natural rubber and therefore provide an approximation on the required magnitude of G' and J_c values and on the ratio of G' values at high and low frequencies. These criteria can be used for the evaluation of the adhesive performance of pressure sensitive adhesives, bearing in mind that individual classes of PSAs will have their unique range of values and properties.

2.2. Comparison of the Adhesive Performance of Silicone and Acrylic PSAs

Both acrylic and silicone PSAs are widely used in the manufacture of transdermal patches. They have good tack and peel properties that tally with the established criteria by Chu and Dahlquist.

The elastic modulus at low frequencies [*G*' (0.1rad/s)] for high tack silicones is around 0.6 x 10^4 Pa, thus slightly lower than Chu's range, whereas for acrylics is around 1.9 x 10^4 Pa, at the lower end of Chu's range.

Oscillatory frequency sweep tests reveal a different behaviour between silicones and acrylics. As shown in figure 1, silicones show viscous behaviour at low frequencies (0.1 rad/s), with their viscous modulus being higher than the elastic modulus (*G*'' > *G*'); and elastic behaviour at high frequencies (100 rad/s), with a predominant elastic modulus (*G*' > *G*''). The high rate of increase of the elastic modulus with frequency leads to an elastic modulus ratio in the middle of Chu's range, at around 150 – 200. In addition the "cross-over" frequency (*G*' = *G*'') can be identified and the relaxation time of the adhesive to be calculated (Ho and Dodou, 2007). As shown in figure 2, acrylics show consistently higher elastic than viscous modulus, with moduli magnitudes increasing almost in parallel across the 0.1 – 100 rad/s frequency range. The slower rate of increase of the elastic modulus with frequency leads to elastic modulus ratios at the lower end of Chu's range, at values of 6 - 7. The difference in viscoelastic behaviour between acrylic and silicone PSAs is also demonstrated by their phase angle, δ, values over the 0.1 – 100 rad/s frequency range; $\delta < 45°$ for acrylics, whereas for silicones $\delta > 45°$ at low frequencies while gradually decreasing to $\leq 45°$ at high frequencies (Figure 3).

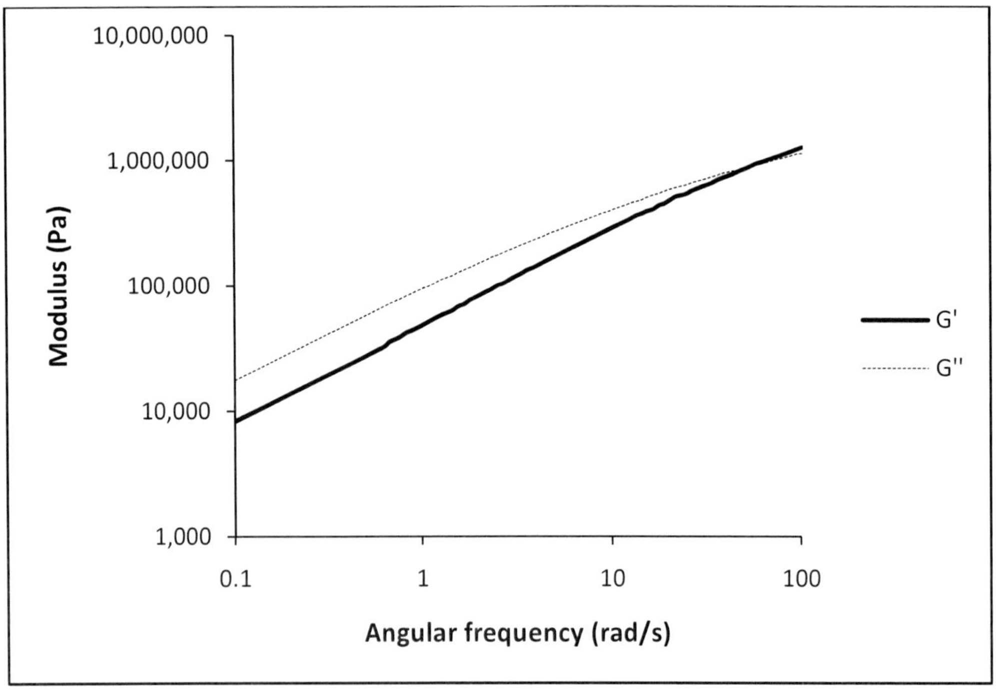

Figure 1. Elastic (G') and viscous (G'') modulus of a silicone adhesive (BIOPSA 4302) over a range of frequencies.

The incorporation of drug and perhaps excipients in the adhesive layer can lead to changes in its viscoelastic behaviour and adhesive performance. Studies on drug-in-adhesive silicone layers have shown a drug concentration-dependent increase on the moduli provided that the drug is dispersed in the adhesive (Ho and Dodou, 2007). When the drug is dissolved in the adhesive, a plasticising effect is often observed leading to decrease in the moduli.

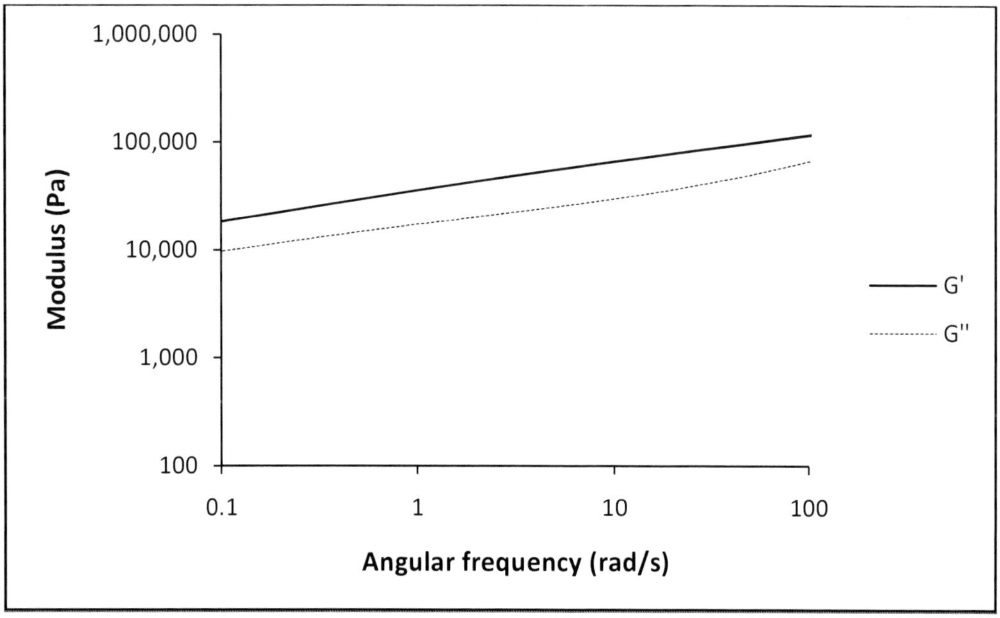

Figure 2. Elastic (G') and viscous (G'') modulus of an acrylic adhesive (DUROTAK 87-4287) versus angular frequency.

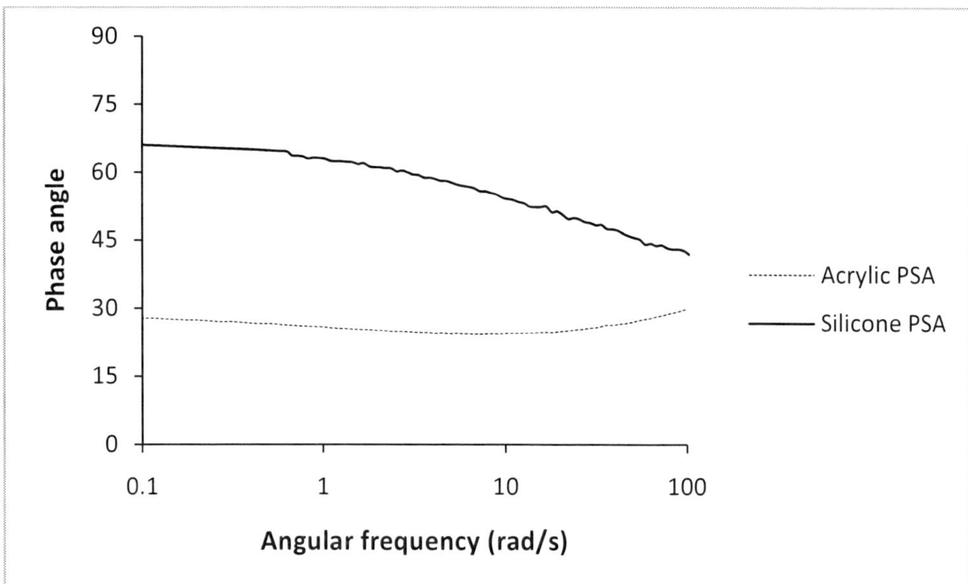

Figure 3. Phase angle (δ) over angular frequency of an acrylic (DUROTAK 87-4287) and silicone (BIOPSA 4302) adhesive.

2.3. Hydrophilic PSAs for Transdermal Application

Hydrophilic adhesive films, in contrast to the traditional hydrophobic PSAs, have also been developed for transdermal drug delivery systems, in an attempt to increase the solubility parameter of the adhesive, thus enabling the incorporation of hydrophilic drugs at high drug loadings and to enhance drug delivery.

Hydrophilic adhesive films for transdermal delivery include: cross-linked blends of hydrophilic polymers, such as PEG-PVP blends (Feldstein et al., 1996); Eudragits, which are cross-linked derivatives of ionic methacrylate copolymers using acetyl tributyl citrate as plasticiser and succinic acid as cross-linking agent; acrylic graft copolymers and silicone graft copolymers, where the hydrophobic PSA monomers have been copolymerised with hydrophilic monomers; and polyurethanes.

All these adhesives are insoluble in water but have the ability to absorb moisture from the skin and swell. Their degree of cross-linking controls their moisture-absorbing capacity and their adhesive-cohesive balance. They can be easily removed from the skin surface by washing but simultaneously they can withstand short contact with water *e.g.* while patient is having a shower, without detaching from the skin. For a detailed description on hydrophilic transdermal adhesives see Feldstein et al, (2009).

3. Hydrophilic Adhesive Films as Mucoadhesive Dosage Forms

Hydrophilic mucoadhesive films are designed to attach on mucosal surfaces. This attachment is accomplished via interaction with mucous.

The mucous lining of internal body cavities is secreted either by specialised cells e.g. goblet cells in the gastrointestinal (GI) tract, or by specialised glands in other mucosal membranes e.g. salivary glands in buccal mucosa. Mucous is a viscoelastic gel matrix and is mainly composed of mucin chains dispersed in 95 – 97% of water. Mucin is a glycoprotein made of a protein backbone and oligosaccharide side chains. The mucous layer varies in thickness in the GI tract; 5 – 500 μm in the stomach, less than 1 μm in the oral cavity. The stomach and duodenal surfaces are covered by a continuous layer of mucous, whereas the intestinal surface is covered intermittently. The mucous layer protects the internal epithelial surfaces from mechanical abrasion e.g. by food, digestive enzymes and acid, and is therefore constantly removed from the membrane and replaced by new mucous. This "turnover" time (about 4 – 5 hours) affects the residence time of the mucoadhesive dosage form and therefore drug bioavailability (Smart, 2005)

Hydrogels are the most popular mucoadhesive polymers. They are hydrophilic polymers that absorb water and swell e.g. chitosan, polyacrylic acid, sodium alginate, cellulose derivatives. Excessive swelling can result in overhydration and dissolution of the adhesive in the mucous with subsequent loss of its adhesive ability. Therefore hydrogels are useful for mucoadhesive dosage forms for administration in cavities with low moisture content e.g. buccal, nasal, ocular, vaginal and rectal cavities.

3.1. Mechanism of Mucoadhesion and Mucoadhesive Polymer Properties

Hydrophilic films when administered are usually in dry state behaving as solids with no adhesive ability. Upon absorption of moisture they become viscoelastic and adhesive. and interact with mucous forming channels or pores within the hydrogel-mucous matrix. Drug then diffuses through the pores and partitions into the epithelial membrane. Drug diffusion rate depends on the polymer properties such as composition, extent of cross-linking and ability to swell, because these properties determine the rate of pore formation within the polymer matrix.

The mucoadhesive bond takes place between the mucoadhesive polymer chains and the mucin chains of the mucous layer. Once the dry polymeric film is positioned in the folds of the mucosal surface, it absorbs moisture from the mucosa and swells to a gel, allowing hydrogel chains to relax and interact with the mucin chains. Mucoadhesive bonds can be: covalent, ionic, Van der Waal's interactions and hydrogen bonds, depending on the nature of the polymer. For example, chitosan forms ionic bonds between its positively charged amino groups and the negatively charged sialic acid groups from mucin.

It is therefore evident that the chemical structure of the polymer determines its ability to form molecular bonds with mucin chains; hydrophilic functional groups, such as hydroxyl groups, carboxylic acids and amines, are required for water uptake, ionic interactions and hydrogen bonding with the mucin chains. In addition, the more similar the chemical structure of the polymer with mucin's chemical structure, the better their miscibility and subsequent interaction. It is therefore desirable for mucoadhesive polymers to have similar solubility parameter with mucin.

The hydration capacity of the polymer is affected both by its molecular weight and degree of cross-linking. The optimum molecular weight ranges between 10,000 Daltons (Da) and 4,000,000 Da. Polymers with too low a molecular weight may overhydrate and dissolve in the mucous, therefore losing their adhesive ability. Whereas polymers with too high a molecular weight will hydrate slowly and not sufficiently enough for their chains to disentangle, leading to poor bonding with mucin. The molecular weight of the polymer is also critical because it determines the length of the polymer chain; long polymer chains will penetrate the mucus and interact better with mucin chains leading to stronger bioadhesion (Smart, 2005).

The degree of cross-linking affects the flexibility of the polymer chains and therefore their ability to move freely. Highly cross-linked polymers will have restricted chain mobility and will not interact sufficiently with the mucin chains. In addition, highly cross-linked polymers have higher moisture absorbing capacity and slower hydration rate; therefore are less likely to overhydrate but simultaneously may not swell promptly enough. Therefore, an optimum degree of cross-linking is desirable.

3.2. Monitoring of Mucoadhesive Performance

Several techniques can be used for the measurement of mucoadhesive performance such as texture analysis (Wong et al., 1999), tensiometry (Varshosaz et al., 2002), tensile strength (Huang et al., 2000; Desai and Kumar, 2004) and rheometry (Tamburic and Graig, 1997; Rossi et al., 1994). The rheological method is based on the phenomenon of synergism, where an efficient interaction between mucoadhesive polymer and mucin leads to increased viscosity of the system, higher than the sum of the individual viscosities of mucous and polymer prior to their interaction (Madsden et al, 1998). The elastic modulus of the system is also increased and this is correlates to the robustness of the mucoadhesive bonding.

Conclusion

The adhesive performance of pressure-sensitive adhesive films for transdermal application and hydrophilic films for transdermal / transmucosal applications is based on their viscoelastic nature and can therefore be correlated with measured viscoelastic parameters. Chu and Dahlquist's established criteria have been used for the adhesive performance evaluation of pressure-sensitive adhesive films. Rheological measurements can also detect differences in the viscoelastic behaviour of different types of adhesive e.g. between acrylic and silicone PSAs, and confer a better insight on their expected tack, peel and "cold flow" properties. Changes in the composition of the film, such as addition of drug and excipients in a drug-in-adhesive design, may affect the adhesive performance of the film and thus monitoring of the adhesive performance is recommended.

Mucoadhesive films are hydrophilic and become viscoelastic upon absorption of moisture from mucus. Rheological measurements of the mucus-mucoadhesive polymer mixture can provide information on the degree of their chain entanglement and bonding.

References

Benedek, I.; Feldstein, M.M. Technology of pressure-sensitive adhesives and products. CRC press Taylor and Francis group: Boca Raton, FL, 2009; chapters 2, 3, 4, 5, 6.

Chu, S-G. In: Adhesive Bonding; Lee, L-H.; Plenum Press: New York, NY, 1991, 97-137.

Dahlquist, C.A. In: Handbook of pressure Sensitive Adhesive Technology; Satas, D.; 3rd Ed.; Satas and Associates: Warwick, RI, 1999; pp 121-138.

Desai, K.G.H.; Kumar, T.M.P. AAPS *PharmSciTech*. 2004, 5(3), article 35.

Feldstein, M.M.; Singh, P.; Cleary, G.W. In: Technology of pressure-sensitive adhesives and products. CRC press Taylor and Francis group: Boca Raton, FL, 2009; 7.1 – 7.79.

Feldstein, M.M.; Tohmakhchi, V.N.; Malkhazov, L.B., Vasiliev A.E., Plate, N.A. *Int. J. Pharm.* 1996, 131, 229-242.

Ho, K.Y.; Dodou, K. *Int. J. Pharm*. 2007, 333, 24-33.

Huang, Y.; Leobandung, W.; Foss, A.; Peppas, N.A. *J. Control. Release*. 2000, 65, 63-71.

Madsden, F.; Eberth, K.; Smart, J.D. *J. Control. Release*. 1998, 50, 167-178.

Muny, R.P. In: Handbook of pressure Sensitive Adhesive Technology; Satas, D.; 3rd Ed.; Satas and Associates: Warwick, RI, 1999; pp 139-152.

Rohn, C.L. In: Handbook of pressure Sensitive Adhesive Technology; Satas, D.; 3rd Ed.; Satas and Associates: Warwick, RI, 1999; pp 153-170.

Rossi, S.; Bonferoni, M.C.; Caramella, C.; Colombo, P. *Eur. J. Pharm. Biopharm*. 1994, 40, 179-182.

Satas, D. Handbook of pressure Sensitive Adhesive Technology. 3rd Ed.; Satas and Associates: Warwick, RI, 1999a; chapters 15, 16, 19, 21, 30.

Satas, D. In: Handbook of pressure Sensitive Adhesive Technology; Satas, D.; 3rd Ed.; Satas and Associates: Warwick, RI, 1999b; pp 36-61.

Smart, J. D. Adv. *Drug Deliv. Reviews*. 2005, 57, 1556-1568.

Sweet R.P.; Ulman, K.L. Healthcare Industries Form; No 51-965A-01; Dow Corning Corporation: Midland, MI, 1997; pp 2-13.

Tamburic, S.; Craig, D.Q.M. *J. Control. Release*. 1995, 37, 59-68.

Tan, H.S.; Pfister, W.R. *PSTT* 1999, 2(2), 60-69.

Ulman, K. L.; Mitchell, T.P.; Loubert, G.L. AAPS, 2006.

Varshosaz, J.; Dehghan, Z. *Eur. J. Pharm. Biopharm*. 2002, 54, 135-141.

Venkatraman, S.; Gale, R. *Biomaterials*. 1998, 19, 1119-1136.

Wokovich, A.M.; Prodduturi, S.; Doub, W.H.; Hussain, A.S.; Buhse, L.F. *Eur. J. Pharm. Biopharm*. 2006, 64, 1-8.

Wong C.F.; Yuen, K.H.; Peh, K.K. *Int. J. Pharm*. 1999, 178, 11-22.

In: Adhesive Properties in Nanomaterials...
Editors: K. A. Wilkinson and D. A. Ordonez
ISBN: 978-1-61209-268-3

Chapter 6

MODELLING ADHESION BY ASYMPTOTIC TECHNIQUES

F. Lebon[1] ***and R. Rizzoni***[2]
[1] Aix-Marseille University and LMNA-GNRS, France
[2] Ferrara University, Italy

Abstract

In this chapter, a review of theoretical and numerical asymptotic studies on thin adhesive layers is proposed. A general mathematical method is presented for modelling the mechanical behavior of bonding and interfaces. This method is based on a simple idea that the adhesive film is supposed to be very thin; the mechanical problem depends strongly on the thinness of the adhesive. It is quite natural, mathematically and mechanically, to consider the limit problem, that is, the asymptotic problem obtained when the thickness and, possibly, the mechanical characteristics of the adhesive thin layer tend to zero. This asymptotic analysis leads to a limit problem with a mechanical constraint on the surface, to which the layer shrinks. The formulation of the limit problem includes the mechanical and geometrical properties of the layer. This limit problem is usually easier to solve numerically by using finite elements software. Theoretical results (i.e. limit problems) can be usually obtained by using at least four mathematical techniques: gamma-convergence, variational analysis, asymptotic expansions and numerical studies. In the chapter, some examples will be presented: comparable rigidity between the adhesive and the adherents, soft interfaces, adhesive governed by a non convex energy and imperfect adhesion between adhesive and adherents. Some numerical examples will also be given and, finally, an example of a numerical algorithm will be presented.

Keywords: Thin layer, interface, asymptotic theory.

1. Introduction

It is now widely admitted that “interphases” (the small volume between two solids) play a crucial role in the analysis of structure assemblies. Nevertheless, due to their small thickness (typically in the 1 μm to 1 mm range), it is difficult in a complete finite element analysis of a structure to take them directly into account. A simplified theory is crucial because of the large number of degrees of freedom and the ill-conditioned numerical problem, which it is very hard to solve even in the linear case. On the other hand, from the numerical point of view, it is of great importance to obtain conditions on the parameters, indicating whenever it is possible to replace the real problem with a simplified one. There exist various strategies to overcome these difficulties. The most classical one is to introduce phenomenological models, taking into account the macroscopic behavior of the interface and eliminating it geometrically. The most famous law of this kind is the classical Coulomb's law [8, 15]. One other possible strategy consists in undertaking an asymptotic analysis (where the small parameter is the thickness of the interphase) to eliminate the interphase geometrically and to obtain an equivalent interface model, which will be simpler to implement in numerical simulations. This idea has been used in many studies and applied, under linear elastic, finite strains, and viscoelastic conditions, to obtain interface laws similar to the phenomenological laws described in the literature [1, 4, 11, 14, 20–27, 31, 33–35, 37–42, 44, 45, 48–50]. Within this approach, the layer no longer exists from the geometrical point of view, but is replaced by a constraint, taking the asymptotic behavior of the parameters into account. In many of these studies, a soft interphase has often been assumed to exist, with a much smaller stiffness than that of the adherents (the stiffness is another small parameter). Fewer studies have focused on joints consisting of adherents and an interphase with a comparable level of rigidity. The present chapter is devoted to a general strategy to analyze this kind of problems.

To summarize, the analysis of thin adhesive bonded joints between deformable bodies involves problems with several parameters. At least two kinds of these parameters are essential because they link the stress vector to the jump in the displacement vector at the interface:

- the thickness of the joint, which is small with respect to those of the adherents,
- possibly, the mechanical characteristics (stiffness, viscosity, etc.) of the joint, which can be smaller than those of the adherents.

To study these asymptotic problems, several mathematical methods have been introduced: matched asymptotic expansions [18, 47], Γ-convergence and variational theory [9, 16], and numerical procedures. After introducing some general notations and the mechanical problem in section 2, we summarize these theories in section 3 and then we apply them to several cases of interphases. In particular, in section 4 we present the asymptotic analysis of a joint made of materials with comparable elastic moduli. Section 5 is devoted to the study of Signorini-Coulomb's conditions at the interfaces between the adherents and the adhesive. In section 6, we report on the study of an interphase characterized by a non convex deformation energy, which is traditionally associated to martensitic phase transformations. An example of the numerical implementation of the laws obtained in section 5 is presented in section 7.

2. General Notations

Let us consider a body occupying an open bounded set Ω of $\mathbb{R}^3$ with a smooth boundary $\partial\Omega$, where the three dimensional space is referred to the orthonormal frame (O, x_1, x_2, x_3).

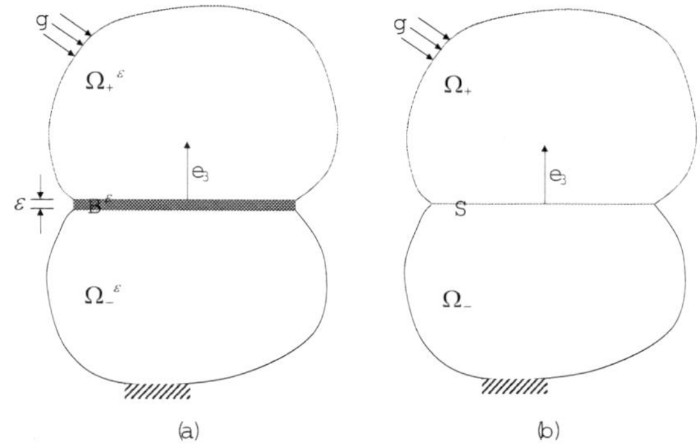

Figure 1. (a) Reference and (b) limit configurations of the joint.

This set Ω is assumed to have a non-empty intersection S with the plane $\{x_3 = 0\}$. Let $\varepsilon > 0$ be a small parameter, and let us define the following sets:

$$\begin{aligned} B^\varepsilon &= \{x = (x_1, x_2, x_3) \in \Omega \,:\, |x_3| < \frac{\varepsilon}{2}\}, \\ \Omega^\varepsilon_\pm &= \{x = (x_1, x_2, x_3) \in \Omega \,:\, \pm\, x_3 > \frac{\varepsilon}{2}\}, \\ \Omega^\varepsilon &= \Omega^\varepsilon_+ \cup \Omega^\varepsilon_-, \\ S^\varepsilon_\pm &= \{x = (x_1, x_2, x_3) \in \Omega \,:\, \pm\, x_3 = \frac{\varepsilon}{2}\}, \\ \Omega_\pm &= \{x = (x_1, x_2, x_3) \in \Omega \,:\, \pm\, x_3 > 0\}, \\ \Omega_0 &= \Omega_+ \cup \Omega_-, \\ S &= \{x = (x_1, x_2, x_3) \in \Omega \,:\, x_3 = 0\}. \end{aligned} \tag{1}$$

B^ε and Ω^ε are the domains occupied by the adhesive and the adherents, respectively (see fig. 1). The structure is subjected to body force density φ and to surface force density g acting on the part Γ_1 of the boundary, whereas it is clamped on the remaining part Γ_0 of the boundary. The adherents and the adhesive are assumed to be linear elastic. We take σ^ε and u^ε to denote the stress tensor and the displacement field, respectively. Under the small perturbation hypothesis, the strain tensor is

$$e_{kh}(u^\varepsilon) = \frac{1}{2}(\frac{\partial u_i^\varepsilon}{\partial x_j} + \frac{\partial u_j^\varepsilon}{\partial x_i}). \tag{2}$$

We take a_{ijkl} to denote the elasticity coefficients of the adherents and λ and μ to stand for the Lamé's coefficients of the glue. We have therefore to solve the following problem

Problem 1

$$
(P_\varepsilon)\begin{cases}
Find\ (u^\varepsilon,\sigma^\varepsilon)\ such\ that: & \\
\sigma^\varepsilon_{ij,j} = -\varphi_i & in\ \Omega \\
\sigma^\varepsilon_{ij} = a_{ijkh}e_{kh}(u^\varepsilon) & in\ \Omega^\varepsilon \\
\sigma^\varepsilon_{ij} = \lambda e_{kk}(u^\varepsilon)\delta_{ij} + 2\mu e_{ij}(u^\varepsilon) & in\ B^\varepsilon \\
u^\varepsilon = 0 & on\ \Gamma_0 \\
\sigma^\varepsilon n = g & on\ \Gamma_1 \\
+\ interface\ laws\ on\ S^\varepsilon_\pm &
\end{cases}
$$

where n denotes the external unit normal vector to Ω. The interface laws between the adherents and the adhesive (perfect interface, Signorini's law or Coulomb's law) will be specified later on.

For a given function $f:\Omega\mapsto R^3$, we define the restrictions of f to the adherents by $f^\pm_\varepsilon$ and to the adhesive by f^m_ε. Denoting $\hat{x}=(x_1,x_2)$ the in-plane coordinates of the adhesive, we define the following jumps of f :

$$
[f]^+_\varepsilon(\hat{x}) := f^+_\varepsilon(x_1,x_2,(\frac{\varepsilon}{2})^+) - f^m_\varepsilon(x_1,x_2,(\frac{\varepsilon}{2})^-)\,, \tag{3}
$$

$$
[f]^-_\varepsilon(\hat{x}) := f^-_\varepsilon(x_1,x_2,(-\frac{\varepsilon}{2})^-) - f^m_\varepsilon(x_1,x_2,(-\frac{\varepsilon}{2})^+)\,, \tag{4}
$$

$$
[f]_\varepsilon(\hat{x}) := f^m_\varepsilon(x_1,x_2,(\frac{\varepsilon}{2})^-) - f^m_\varepsilon(x_1,x_2,(-\frac{\varepsilon}{2})^+)\,. \tag{5}
$$

For a given function $f:\Omega_0\mapsto R^3$, we define the restrictions of f to $\Omega_\pm$ by $f^\pm$ and we also define the following jump of f on S :

$$
[f](\hat{x}) := f^+(x_1,x_2,0^+) - f^-(x_1,x_2,0^-)\,. \tag{6}
$$

Finally,we also introduce the following notations:

$$
[[f]]^\pm_{N\varepsilon} := \pm[f]^\pm_\varepsilon e_3, \tag{7}
$$

$$
[[f]]^\pm_{T\varepsilon} := [f]^\pm_\varepsilon \mp [[f]]^\pm_{N\varepsilon} e_3, \tag{8}
$$

$$
\sigma^{\varepsilon\pm}_N := \pm(\sigma^\varepsilon e_3)^\pm_\varepsilon e_3, \tag{9}
$$

$$
\sigma^{\varepsilon\pm}_T := \pm\sigma^\varepsilon e_3 \mp \sigma^{\varepsilon\pm}_N e_3. \tag{10}
$$

3. Mathematical Methods

3.1. Matched Asymptotic Expansion Method

The idea underlying matched asymptotic expansions is to find two expansions of the displacement u^ε and the stress σ^ε in powers of ε, i.e., an external expansion for the bodies and an internal one for the joint, and to connect these two expansions to obtain the same limit [18,47].

3.1.1. External Expansions

The external expansion is a classical expansion in powers of ε

$$\begin{array}{rcl} u^{\varepsilon}(\hat{x},x_3) & = & u^0(\hat{x},x_3)+\varepsilon u^1(\hat{x},x_3)+...,\\ e_{ij}(u^{\varepsilon})(\hat{x},x_3) & = & e^0_{ij}(\hat{x},x_3)+\varepsilon e^1_{ij}(\hat{x},x_3)+...,\\ e^l_{ij} & = & \frac{1}{2}(\frac{\partial u^l_i}{\partial x_j}+\frac{\partial u^l_j}{\partial x_i}),\\ \sigma^{\varepsilon}_{ij}(\hat{x},x_3) & = & \sigma^0_{ij}(\hat{x},x_3)+\varepsilon\sigma^1_{ij}(\hat{x},x_3)+... \end{array} \tag{11}$$

3.1.2. Internal Expansions

In the internal expansions, we perform a blow-up of the second variable. Let $y_3 = \frac{x_3}{\varepsilon}$. The internal expansion gives

$$\begin{array}{rcl} u^{\varepsilon}(\hat{x},y_3) & = & v^0(\hat{x},y_3)+\varepsilon v^1(\hat{x},y_3)+...,\\ e_{ij}(u^{\varepsilon})(\hat{x},y_3) & = & \varepsilon^{-1}e^{-1}_{ij}(\hat{x},y_3)+e^0_{ij}(\hat{x},y_3)+\varepsilon e^1_{ij}(\hat{x},y_3)+...,\\ e^l_{\alpha\alpha} & = & \frac{\partial v^l_\alpha}{\partial x_\alpha},\quad \alpha=1,2,\\ e^l_{33} & = & \frac{\partial v^{l+1}_3}{\partial y_3},\\ e^l_{\alpha\beta} & = & \frac{1}{2}(\frac{\partial v^l_\beta}{\partial x_\alpha}+\frac{\partial v^l_\alpha}{\partial y_\beta}),\\ e^l_{\alpha 3} & = & \frac{1}{2}(\frac{\partial v^l_3}{\partial x_\alpha}+\frac{\partial v^{l+1}_\alpha}{\partial y_3}),\\ \sigma^{\varepsilon}_{ij}(\hat{x},y_3) & = & \varepsilon^{-1}\tau^{-1}_{ij}(\hat{x},y_3)+\tau^0_{ij}(\hat{x},y_3)+\varepsilon\tau^1_{ij}(\hat{x},y_3)+...,\\ \sigma^{\varepsilon}_{ij,j} & = & \sum_{l=-2}^{\infty}\varepsilon^l(\frac{\partial\tau^l_{i\alpha}}{\partial x_\alpha}+\frac{\partial\tau^{l+1}_{i3}}{\partial y_3}). \end{array} \tag{12}$$

In the last relation, we use the convention that $v^l = 0$ when $l < 0$ and $\tau^l = 0$ when $l < -1$.

3.1.3. Continuity Conditions

The third step of the method consists in connecting the two expansions. We choose some intermediate planes defined by $x_3 = \pm\zeta\varepsilon^t$, $0 < t < 1$, $\zeta \in]0, +\infty[$. When ε tends to zero, x_3 tends to $0^{\pm}$ and $y_3 = x_3/\varepsilon$ tends to $\pm\infty$. The principle of the method [18, 47] consists in assuming that the two expansions give both the same asymptotic limits, that is

$$\begin{array}{rcl} (i)\ v^0(\hat{x},\pm\infty) & = & u^0(\hat{x},0^{\pm}),\\ (ii)\ \tau^{-1}(\hat{x},\pm\infty) & = & 0,\\ (iii)\ \tau^0(\hat{x},\pm\infty) & = & \sigma^0(\hat{x},0^{\pm}). \end{array} \tag{13}$$

3.1.4. Identification

The fourth step of the method consists in introducing the expansions in the equilibrium equations and in the constitutive equations. We obtain systems of equations in the following form:

$$\sum \varepsilon^n B^n(v^l, \tau^l) = 0 \tag{14}$$

We now impose that these relations are satisfied at each order n :

$$B^n(.) = 0, \ \forall\, n. \tag{15}$$

These equations together with (13) identify an interface law [33, 34].

3.2. Γ-convergence Theory

When Problem 1 admits an energy form, it is natural to study the variational convergence of this energy F^ε, as the small parameter ε tends to zero. A now classical theory to analyze the convergence is provided by the notion of $\Gamma-$ convergence [9, 16].

We recall the following definition:

Definition 1 *Let X be a topological space and $F^\varepsilon : X \to [0,\infty]$ a sequence of functionals on X. The sequence F^ε is said to Γ-converge to the Γ-limit $F^0 : X \to [0,\infty]$ if the following two conditions hold:*

- *For every sequence u^ε in X such that $u^\varepsilon \to u^0$ as $\varepsilon \to 0$,*
  $F^0(u^0) \leq \liminf_{\varepsilon\to 0} F^\varepsilon(u^\varepsilon)$ (Lower bound inequality);
- *For every $u^0 \in X$, there exists a sequence u^ε converging to u^0 as $\varepsilon \to 0$, such that*
  $F^0(u^0) \geq \limsup_{\varepsilon\to 0} F^\varepsilon(u^\varepsilon)$ (Upper bound inequality).

F^0 is an optimal asymptotic lower bound for the sequence F^ε in the sense specified by the following proposition.

Proposition 2 *If F^ε $\Gamma-$converges to F^0, then*

$$\text{(i)}\ \ F^\varepsilon(u^\varepsilon) := \min_{u\in X} F^\varepsilon(u) \to F^0(u^0) := \min_{u\in X} F^0(u), \tag{16}$$

$$\text{(ii)}\ u^\varepsilon \to u^0. \tag{17}$$

This theory, applied to the total energy of the system, allows to obtain an interface law at order zero [36, 42].

3.3. Energy Asymptotical Method

This method has been introduced in [36]. As in the $\Gamma-$ convergence theory, the equilibrium problem is written as a minimization problem of the total energy. As in the asymptotic expansion method, the energy asymptotical method is based on an (external) expansion of u in power of ε. Moreover, in order to reformulate the equilibrium problem in an interphase domain independent of ε, a change of variables is introduced:

$$(\hat{z}, z_3) \ = \ (\hat{x}, x_3\varepsilon^{-1}), \qquad (\hat{x}, x_3) \in B_\varepsilon, \tag{18}$$

$$(\hat{z}, z_3) \ = \ (\hat{x}, x_3 \pm \frac{\varepsilon}{2} \mp \frac{1}{2}), \ \ (\hat{x}, x_3) \in \Omega_\pm. \tag{19}$$

In particular, B_ε is rescaled by a factor ε^{-1} along the interphase thickness and the bodies $\Omega_\pm$ are shifted by $\pm 1/2(1-\varepsilon)$ in the same direction.

Let $\bar{u}^\varepsilon(\hat{z},z_3)$ be the rescaled displacement. The rescaled equilibrium problem can be formulated as follows: find the $\bar{u}^\varepsilon$ minimizing the rescaled energy, $\mathcal{F}^\varepsilon$, in the set of rescaled displacements. Substituting the expansion $(11)_1$ into the rescaled energy, we obtain

$$\mathcal{F}^\varepsilon(u^l) = \frac{1}{\varepsilon}\mathcal{F}^{-1}(u^l) + \mathcal{F}^0(u^l) + \varepsilon\mathcal{F}^1(u^l) + \varepsilon^2\mathcal{F}^2(u^l) + o(\varepsilon^2) \tag{20}$$

The main assumption of this method is that we can obtain the fields which are stationary points of the energy $\mathcal{F}^\varepsilon$ by finding the stationary points of the energies $\mathcal{F}^l$ obtained at each level of the expansion.

3.4. Numerical Methods

In this part, we present a method used in previous studies. Examples, usually academic, are studied numerically (long bar, dovetail assembly, etc.). The aim of this numerical study by finite elements is to confirm that the numerical results are coherent with a theoretical approach, to check the existence of the theory (validation) and to determine the thickness of the layer at which the limit interface law can be taken to be a valid approximation of the original constitutive relation of the adhesive (quantification). The initial problem is therefore solved numerically in the case of a thin layer with decreasing thickness (and decreasing moduli in the case of a soft interphase). The numerical results obtained are then compared with the theoretical results. For example, in the case of a soft elastic isotropic interphase, in order to simplify the computations, the Lamé's coefficients are assumed to be given by

$$\lambda_\varepsilon = \lambda^\delta, \mu_\varepsilon = \mu^\gamma$$

where λ, μ are given and the non negative coefficients δ, γ correspond to cases of soft or rigid interfaces. We give an example in section 7, where the problem is solved with the following decreasing values of ε: $1.10^{-2}, 5.10^{-3}, 1.10^{-3}, 5.10^{-4}, 1.10^{-4}$. This method is used in [31,39].

4. The Case Involving Similar Rigidity: Perfect Interface?

In this section, we consider:

- that the adhesive is linear elastic,
- that the link between the adhesive and the adherents is perfect,
- that the stiffness of the adhesive is similar to those of the adherents.

We present the results obtained in [35], where we have used $\Gamma-$convergence techniques to analyze the asymptotic behavior of the total energy up to the second order of the expansion in ε.

4.1. First Order Study

The local formulation of the equilibrium problem corresponds to find the fields of displacement u^ε, strain $e(u^\varepsilon)$ and stress σ^ε which solve the following problem:

$$(P_\varepsilon^1)\left\{\begin{array}{ll} Find\ (u^\varepsilon,\sigma^\varepsilon)\ such\ that: & \\ \sigma^\varepsilon_{ij,j}=-\varphi_i & in\,\Omega \\ \sigma^\varepsilon_{ij}=a^{\pm}_{ijkh}e_{kh}(u^\varepsilon) & in\,\Omega^\varepsilon_\pm \\ \sigma^\varepsilon_{ij}=a^{m}_{ijkh}e_{kh}(u^\varepsilon) & in\,B^\varepsilon \\ u^\varepsilon=0 & on\,\Gamma_0 \\ \sigma_\varepsilon n=g & on\,\Gamma_1 \\ [u^\varepsilon]^\pm_\varepsilon=0\,,\ [\sigma^\varepsilon e_3]^\pm_\varepsilon=0 & \end{array}\right.$$

We introduce the space of admissible displacements (the Sobolev spaces are denoted by usual notations)

$$V^\varepsilon=\{u\in(W^{1,2}(\Omega))^3: u=0\ \text{ on}\,\Gamma_0\}\,, \tag{21}$$

and the works of the internal and external loads, respectively,

$$\begin{aligned} A^\varepsilon(u,v) &= \int_{\Omega^\varepsilon_\pm} a^\pm e(u):e(v)\,dx+\int_{B^\varepsilon} a^m e(u):e(v)\,dx\,, \\ l^\varepsilon(v) &= \int_\Omega \phi\cdot v\,dx+\int_{\Gamma_1} g\cdot v\,ds\,, \end{aligned}$$

we reformulate (P_ε^1) in the following way

$$(\bar P_\varepsilon^1)\ \left\{\begin{array}{l}\text{Find }\ u^\varepsilon\in V_\varepsilon: \\ A(u^\varepsilon,v)=l^\varepsilon(v)\ \ \forall v\in V^\varepsilon\,.\end{array}\right. \tag{22}$$

Under the three following regularity assumptions

$$H1)\qquad \left\{\begin{array}{l} a_{ijkl}\in L^\infty(\Omega)\,, \\ a_{ijkl}=a_{klij}=a_{jilk} \\ \exists\eta>0: a_{ijkl}e_{ij}e_{kl}\ge\eta e_{ij}e_{ij}\ \ \forall e_{ij}=e_{ji}\,, \end{array}\right. \tag{23}$$

$$H2)\qquad \exists\varepsilon_0: B_\varepsilon\cap(\Gamma_1\cup\text{supp}(\phi))=\emptyset\,,\ \ \forall\ \varepsilon<\varepsilon_0\,, \tag{24}$$

$$H3)\qquad \phi\in(L^2(\Omega))^3\,,g\in(L^2(\Gamma_1))^3\,.$$

the variational problem $(\bar P_\varepsilon^1)$ admits a unique solution (due to the Lax-Milgram theorem), which is also the unique solution of the following problem of minimization:

$$(\tilde P_\varepsilon^1)\ \left\{\begin{array}{l}\text{Find }\ u^\varepsilon\in V^\varepsilon: \\ J^\varepsilon(u^\varepsilon)\le J^\varepsilon(v)\ \ \forall v\in V^\varepsilon\,,\end{array}\right. \tag{25}$$

where

$$J^\varepsilon(v)=\frac{1}{2}A^\varepsilon(v,v)-l^\varepsilon(v) \tag{26}$$

is the potential energy associated with the displacement field v. To apply the notion of $\Gamma-$convergence, we introduce the strain energy functional

$$F^{\varepsilon}(v) = \begin{cases} \frac{1}{2}A^{\varepsilon}(v,v) & \text{if } v \in V_{\varepsilon} , \\ +\infty & \text{if } v \in X \setminus V_{\varepsilon} , \end{cases} \tag{27}$$

where $X = (L^2(\Omega))^3$. In [35], it is introduced the following limit functional:

$$F^0(v) = \begin{cases} \int_{\Omega_0} a^{\pm} e(v) : e(v)\, dx & \text{if } v \in V^0 , \\ +\infty & \text{if } v \in X \setminus V^0 , \end{cases} \tag{28}$$

where

$$V^0 = \{u \in (W^{1,2}(\Omega_0))^3 : u = 0 \text{ on } \Gamma_0 ,\ [v] = 0 \text{ on } S\} . \tag{29}$$

We now recall the following result proved in [35].

Theorem 3 *The sequence of functionals F^{ε} Γ-converges to F^0 in the strong topology of X.*

Note that the limit problem, i. e. the minimization problem of F^0, involves only the adherents. The adhesive has vanished geometrically (as expected) as well as mechanically. We also observe that, because the displacement vector on S has to be continuous (see (29)), the mechanical implication of the theorem is that a very thin interphase behaves like a perfect interface.

In some cases, it may become necessary to improve the model, and it turns out that the most natural solution is to go to a higher order in the asymptotic analysis. This is done in the next section.

4.2. Second Order Study

In this section, we recall the results obtained in [35] at the second order and we focus on the case of isotropic adhesive with Lamé's coefficients λ and μ. We introduce the displacement field u^1 and the stress field σ^1 as the limit, in a sense made precise by the following lemma, of suitable subsequences of $\frac{u^{\varepsilon} - u^0}{\varepsilon}$ and $\frac{\sigma^{\varepsilon} - \sigma^0}{\varepsilon}$ respectively. We denote $D(A)$ the space of the C^{∞} functions with compact support on the open set A and $D'(A)$ its dual space.

Lemma 1 *Let u^{ε}, σ^{ε} and u^0, and σ^0 be the displacement and stress fields corresponding to minimizers of the energies F^{ε} and F^0, respectively. Then there exist subsequences, not relabeled, such that*

$$\begin{aligned} \frac{u^{\varepsilon} - u^0}{\varepsilon} &\rightharpoonup u^1 \text{ in } L^2(\Omega)\ (weak), \\ \frac{\sigma^{\varepsilon} - \sigma^0}{\varepsilon} &\rightharpoonup \sigma^1 \text{ in } D'(\Omega)\ (weak). \end{aligned} \tag{30}$$

This Lemma is proved in [35]. Contrarily to the fields at order zero u^0 and $\sigma^0 e_3$ which are continuous on the surface S, the fields u^1 and $\sigma^1 e_3$ suffer discontinuity on S. The following Theorem gives a relationship between their jumps, involving the restrictions of the traction and the displacement vectors u^0 and $\sigma^0 e_3$ on S.

Theorem 4 *The fields u^1 and σ^1 satisfy (in a weak sense) the following equilibrium equations:*

$$\begin{aligned} \sigma^1_{ij,j} &= 0 \quad in\, \Omega_\pm \\ \sigma^1_{ij} &= a^\pm_{ijkh} e_{kh}(u^1) \quad in\, \Omega_\pm \\ u^1 &= 0 \quad on\, \Gamma_0 \\ \sigma^1 n &= 0 \quad on\, \Gamma_1 \end{aligned}$$

and the following relations at the interface S :

$$\begin{aligned} [u^1_\alpha] &= \frac{1}{\mu}\sigma^0_{\alpha 3}(\hat{x},0) - u^0_{3,\alpha}(\hat{x},0) - \frac{1}{2}(u^0_{\alpha,3}(\hat{x},0^+) + u^0_{\alpha,3}(\hat{x},0^-)) , \quad \alpha = 1,2 , \\ [u^1_3] &= \frac{1}{\lambda+2\mu}\sigma^0_{33}(\hat{x},0) - \frac{\lambda}{\lambda+2\mu}(u^0_{1,1}(\hat{x},0) + u^0_{2,2}(\hat{x},0)) \\ &\quad -\frac{1}{2}(u^0_{3,3}(\hat{x},0^+) + u^0_{3,3}(\hat{x},0^-)) , \\ [\sigma^1_{13}] &= -\frac{4\mu(\lambda+\mu)}{\lambda+2\mu}u^0_{1,11}(\hat{x},0) - \mu u^0_{1,22}(\hat{x},0) - \frac{\mu(3\lambda+2\mu)}{\lambda+2\mu}u^0_{2,21} \\ &\quad -\frac{\lambda}{\lambda+2\mu}\sigma^0_{33,1}(\hat{x},0) - \frac{1}{2}(\sigma^0_{13,3}(\hat{x},0^+) + \sigma^0_{13,3}(\hat{x},0^-)) , \\ [\sigma^1_{23}] &= -\frac{4\mu(\lambda+\mu)}{\lambda+2\mu}u^0_{2,22}(\hat{x},0) - \mu u^0_{2,11}(\hat{x},0) - \frac{\mu(3\lambda+2\mu)}{\lambda+2\mu}u^0_{1,12}(\hat{x},0) \\ &\quad -\frac{\lambda}{\lambda+2\mu}\sigma^0_{33,2}(\hat{x},0) - \frac{1}{2}(\sigma^0_{23,3}(\hat{x},0^+) + \sigma^0_{23,3}(\hat{x},0^-)) , \\ [\sigma^1_{33}] &= -\sigma^0_{13,1}(\hat{x},0) - \sigma^0_{23,2}(\hat{x},0) - \frac{1}{2}(\sigma^0_{33,3}(\hat{x},0^+) + \sigma^0_{33,3}(\hat{x},0^-)) . \end{aligned} \tag{31}$$

As for the minimization problem of the energy (28), the limit problem defined by theorem 4 involves only the adherents, since the adhesive has vanished geometrically. However, the adhesive has not disappeared from the mechanical point of view: it has been replaced by the mechanical constraints (31), linking the jump in the displacement and traction vectors to the displacement and traction vectors (and their derivatives) at order zero. In particular, the presence of the tangential derivatives of u_0 indicates a non-local character of the interface law (31). Relations (31) also suggest that at higher orders, a thin interphase behaves like an imperfect interface, which keeps trace of the elastic behavior of the interphase.

4.3. An One Dimensional Example

In this section, we propose a simple example to illustrate, in the one dimensional setting, the origin of the terms entering the interface law (31). We consider a bar, AB, fixed at one of its extremities, A, as represented in Figure 2. The part AB is composed of the adhesive AC and of the adherent CB, made by two different materials with elastic moduli E_1 and E_2, respectively. A force density $\phi(x) = g_1 x$ is applied on AC and a force density $\phi(x) = g_2 x$ is applied on CB. We have

$$AC = \varepsilon,\ CD = L - \varepsilon,$$

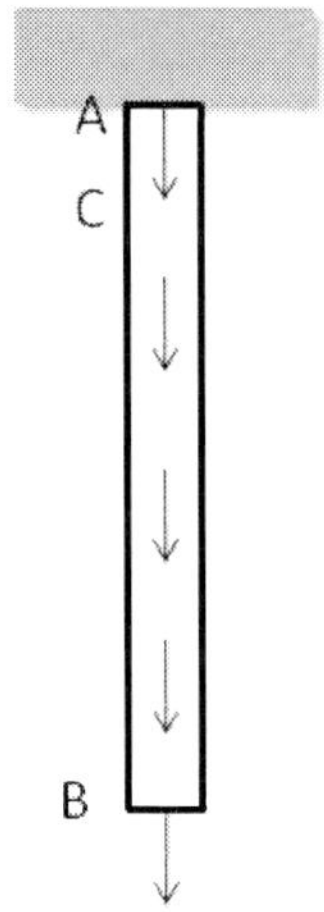

Figure 2. A composite bar subject to body forces.

$$u(0) = 0, \; E_2 \frac{du}{dx}(L) = a.$$

At order zero, we obtain:

$$u^0(x) = -\frac{g_2 x^3}{6E_2} + \frac{(a - 1/2 g_2 L^2)x}{E_2} \quad (32)$$

$$\sigma^0(x) = -\frac{g_2}{2}x^2 + a - 1/2 g_2 L^2 \quad (33)$$

We verify easily that $[u^0] = [\sigma^0] = 0$. At order one, we obtain:

$$u^1(x) = (\frac{1}{E_1} - \frac{1}{E_2})(a - 1/2 g_2 L^2) \quad (34)$$

$$\sigma^1(x) = 0. \quad (35)$$

We can verify that, $[u^1] = \frac{1}{E_1}\sigma^0(0) - u^0_{,x}(0)$, and $[\sigma^1] = -\frac{1}{2}(\sigma^0_{,x}(0^+) + \sigma^0_{,x}(0^-))$.

5. Non Linear Imperfect Interface: Taking into Account Frictional Contact

In this section, we consider:

- that the thin layer is elastic,
- that the stiffness of the glue is lower than that of the bodies,
- that the link between the glue and the bodies is imperfect.

We assume the contact to involve dry friction conditions between the bodies and the thin layer. The Signorini's law of unilateral contact and the Coulomb's law of dry friction are written in the case of monotonous quasi-static loading as:

$$\begin{array}{rcl} [\sigma^{\varepsilon}e_3]^{\pm}_{\varepsilon} & = & 0, \\ \sigma^{\varepsilon\pm}_{N} & \leq & 0, \\ [[u^{\varepsilon}]]^{\pm}_{N\varepsilon} & \leq & 0, \\ \sigma^{\varepsilon\pm}_{N}[[u^{\varepsilon}]]^{\pm}_{N\varepsilon} & = & 0, \\ |\sigma^{\varepsilon\pm}_{T}| & \leq & f|\sigma^{\varepsilon\pm}_{N}|, \\ If\ |\sigma^{\varepsilon\pm}_{T}| < f|\sigma^{\varepsilon\pm}_{N}| & then & [[u^{\varepsilon}]]^{\pm}_{T\varepsilon} = 0, \\ If\ |\sigma^{\varepsilon\pm}_{T}| = f|\sigma^{\varepsilon\pm}_{N}| & then & [[u^{\varepsilon}]]^{\pm}_{T\varepsilon} = -\zeta\sigma_T,\ with\ \zeta \geq 0, \end{array} \quad (36)$$

where f is the friction coefficient. To analyze the asymptotic behavior of the interphase, we use two of the methods presented in section 3: matched asymptotic expansion and a numerical method.

5.1. Asymptotic Expansions

In the following, we summarize the results obtained in [39]. Given the constitutive equations in the thin layer, we have

$$\begin{array}{rcl} \varepsilon^{-1}\tau^{-1}_{ij} + \tau^{0}_{ij} + \varepsilon\tau^{1}_{ij} + ... & = & \lambda(\varepsilon^{-1}e^{-1}_{kk} + e^{0}_{kk} + \varepsilon e^{1}_{kk} + ...)\delta_{ij} \\ & & +2\mu(\varepsilon^{-1}e^{-1}_{ij} + e^{0}_{ij} + \varepsilon e^{1}_{ij} + ...). \end{array} \quad (37)$$

The identification of the various orders depends on the behavior of the Lamé coefficients λ and μ with respect to ε. We obtain nine cases corresponding to the values of the limits $\overline{\lambda}$ and $\overline{\mu}$ of the ratios $\frac{\lambda}{\varepsilon}$ and $\frac{\mu}{\varepsilon}$:

- (a) $\overline{\lambda} = \overline{\mu} = 0,\ \tau^{-1}_{ij} = 0,\ \tau^{0}_{ij} = 0.$
- (b) $\overline{\lambda} = 0,\ 0 < \overline{\mu} < \infty,\ \tau^{-1}_{ij} = 0,\ \tau^{0}_{ij} = 2\overline{\mu}e^{-1}_{ij}.$
- (c) $0 < \overline{\lambda} < \infty,\ \overline{\mu} = 0,\ \tau^{-1}_{ij} = 0,\ \tau^{0}_{ij} = \overline{\lambda}e^{-1}_{kk}\delta_{ij}.$
- (d) $0 < \overline{\lambda} < \infty,\ 0 < \overline{\mu} < \infty,\ \tau^{-1}_{ij} = 0,\ \tau^{0}_{ij} = \overline{\lambda}e^{-1}_{kk}\delta_{ij} + 2\overline{\mu}e^{-1}_{ij},$
- (e) $\overline{\lambda} = \infty,\ \overline{\mu} = 0,\ e^{-1}_{kk} = 0,\ \tau^{0}_{ij} = 0.$
- (f) $\overline{\lambda} = \infty,\ 0 < \overline{\mu} < \infty,\ e^{-1}_{kk} = 0,\ \tau^{0}_{ij} = 2\overline{\mu}e^{-1}_{ij}.$
- (g) $\overline{\lambda} = 0,\ \overline{\mu} = \infty,\ e^{-1}_{ij} = 0,\ e^{0}_{ij} = 0.$
- (h) $0 < \overline{\lambda} < \infty,\ \overline{\mu} = \infty,\ e^{-1}_{ij} = 0,\ e^{0}_{ij} = 0.$
- (i) $\overline{\lambda} = \infty,\ \overline{\mu} = \infty,\ e^{-1}_{ij} = 0,\ e^{0}_{ij} = 0.$

Let $K_1 = K_T = \overline{\mu}$ and $K_2 = K_N = \overline{\lambda} + 2\overline{\mu}$. f^+ is the positive part of a function f.

- (a) $\overline{\lambda} = \overline{\mu} = 0$
  In this case, $\tau^0_{ij} = 0$. Because of the continuity conditions, we obtain

$$\sigma^0(x_1, 0) = 0 \tag{38}$$

- (b) $\overline{\lambda} = 0, 0 < \overline{\mu} < \infty$
  In this case, $\tau^0_{11} = 0, \tau^0_{12} = \overline{\mu}\frac{\partial v^0_1}{\partial y_2}, \tau^0_{22} = 2\overline{\mu}\frac{\partial v^0_2}{\partial y_2}$.
  Using standard arguments, we obtain

$$\begin{array}{rcl} \sigma^0_N & = & -2\overline{\mu}[[u^0]]^+_N, \\ |\sigma^0_T| & \leq & f|\sigma^0_N|, \\ If\ |\sigma^0_T| < f|\sigma^0_N| & then & \sigma^0_T = -\overline{\mu}[[u^0]]_T, \\ If\ |\sigma^0_T| = f|\sigma^0_N| & then & [[u^0]]_T = -\zeta\sigma^0_T, \zeta \geq 0. \end{array} \tag{39}$$

- (c) $0 < \overline{\lambda} < \infty, \overline{\mu} = 0$
  In this case $\tau^0_{11} = \tau^0_{22} = \overline{\lambda}\frac{\partial v^0_2}{\partial y_2}, \tau^0_{12} = 0$. For the normal part, we proceed as in (b), by replacing $2\overline{\mu}$ by $\overline{\lambda}$. The tangential part corresponds to (a). We obtain

$$\begin{array}{rcl} \sigma^0_N & = & -\overline{\lambda}[[u^0]]^+_N, \\ \sigma^0_T & = & 0. \end{array} \tag{40}$$

- (d) $0 < \overline{\lambda} < \infty, 0 < \overline{\mu} < \infty$
  In this case, $\tau^0_{11} = \overline{\lambda}\frac{\partial v^0_2}{\partial y_2}, \tau^0_{i2} = K_i\frac{\partial v^0_i}{\partial y_2}$. Here, we again proceed as in (b). We obtain

$$\begin{array}{rcl} \sigma^0_N & = & -(\overline{\lambda} + 2\overline{\mu})[[u^0]]^+_N, \\ |\sigma^0_T| & \leq & f|\sigma^0_N|, \\ If\ |\sigma^0_T| < f|\sigma^0_N| & then & \sigma^0_T = -\overline{\mu}[[u^0]]_T, \\ If\ |\sigma^0_T| = f|\sigma^0_N| & then & [[u^0_]]T = -\zeta\sigma^0_T, \zeta \geq 0. \end{array} \tag{41}$$

- (e) $\overline{\lambda} = \infty, \overline{\mu} = 0$
  We have $\frac{\partial v^0_2}{\partial y_2} = 0, \tau^0_{12} = 0$. For the normal tangential part, we proceed as in (a). Using standard arguments, we have

$$\begin{array}{c} [[u^0]]_N \leq 0, \quad \sigma^0_N \leq 0, \quad [[u^0]]_N\sigma^0_N = 0, \\ \sigma^0_T \quad = \quad 0. \end{array} \tag{42}$$

- (f) $\overline{\lambda} = \infty, 0 < \overline{\mu} < \infty$
  We have $\frac{\partial v^0_2}{\partial y_2} = 0, \tau^0_{12} = \overline{\mu}\frac{\partial v^0_1}{\partial y_2}$. The treatment of the tangential part is similar to (b) and the treatment of the normal part is similar to (e). To summarize, we obtain

$$\begin{array}{rcl} [[u^0]]_N \leq 0, \quad \sigma^0_N \leq 0, & & [[u^0]]_N\sigma^0_N = 0, \\ |\sigma^0_T| & \leq & f|\sigma^0_N|, \\ If\ |\sigma^0_T| < f|\sigma^0_N| & then & \sigma^0_T = -\overline{\mu}[[u^0]]_T, \\ If\ |\sigma^0_T| = f|\sigma^0_N| & then & [[u^0]]_T = -\zeta\sigma^0_T, \zeta \geq 0. \end{array} \tag{43}$$

The three last cases call for a special treatment.

- (g)(h)(i) $\overline{\mu} = \infty$

  In these cases, we have $\dfrac{\partial v_i^0}{\partial y_2} = 0$. We proceed as in (e) on the normal and tangential parts. This gives

$$\begin{array}{rcl} [[u^0]]_N \leq 0, & \sigma_N^0 \leq 0, & [[u^0]]_N \sigma_N^0 = 0, \\ |\sigma_T^0| & \leq & f|\sigma_N^0|, \\ If\ |\sigma_T^0| < f|\sigma_N^0| & then & [[u^0]]_T = 0, \\ If\ |\sigma_T^0| = f|\sigma_N^0| & then & [[u^0]]_T = -\zeta\sigma_T^0, \zeta \geq 0. \end{array} \tag{44}$$

5.2. Numerical Validations

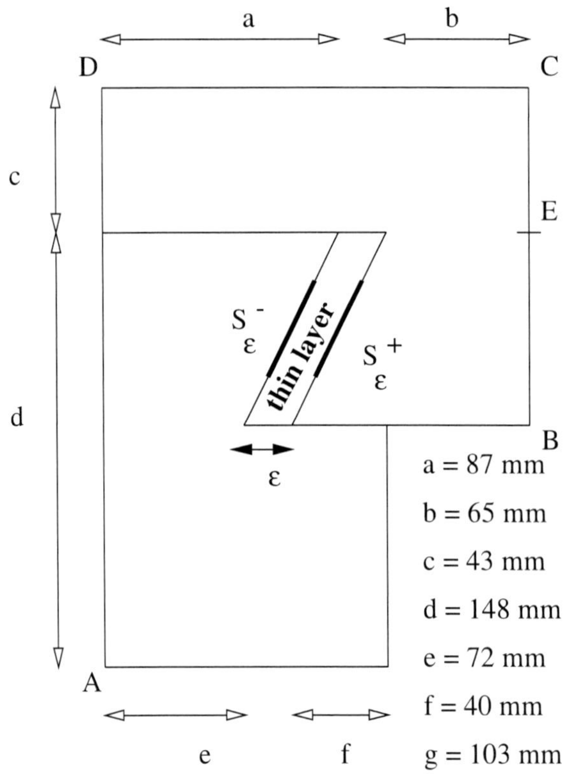

Figure 3. The dovetail assembly.

This section illustrates some numerical results obtained in [31, 44]. The example of a dovetail assembly is treated. The aim of this numerical study is to confirm that the numerical results are coherent with the theory (validation) and to determine the thickness of the layer at which the limit interface law can be taken to be valid (quantification). The initial problem is therefore solved numerically in the case of a thin layer with decreasing thickness and stiffness values, as done by [44]. A relaxation procedure is used for this purpose, as in previous studies [30, 46].

The numerical results obtained are then compared with the theoretical results. In order to

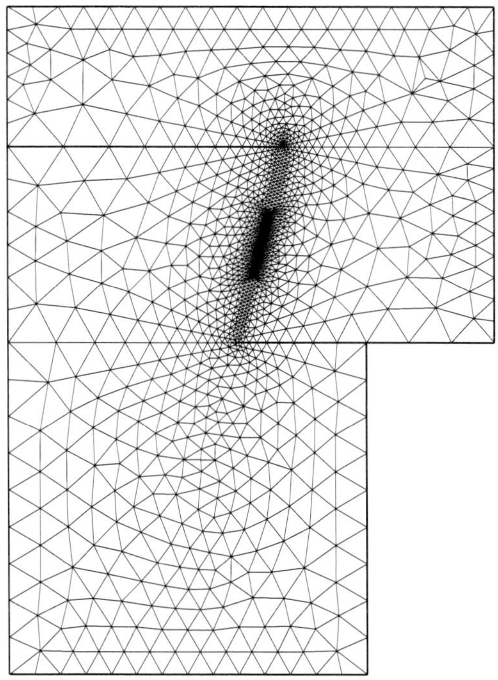

Figure 4. Mesh of the structure.

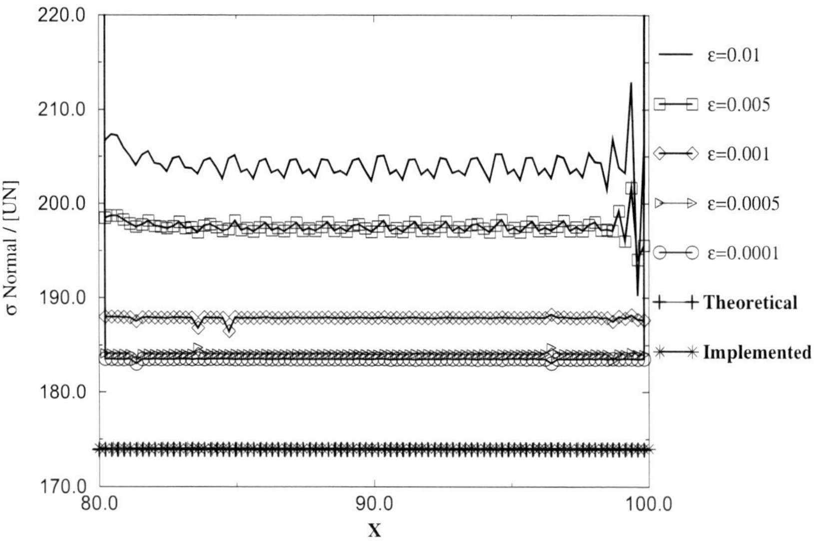

Figure 5. Ratio between normal stress and normal displacement (case b).

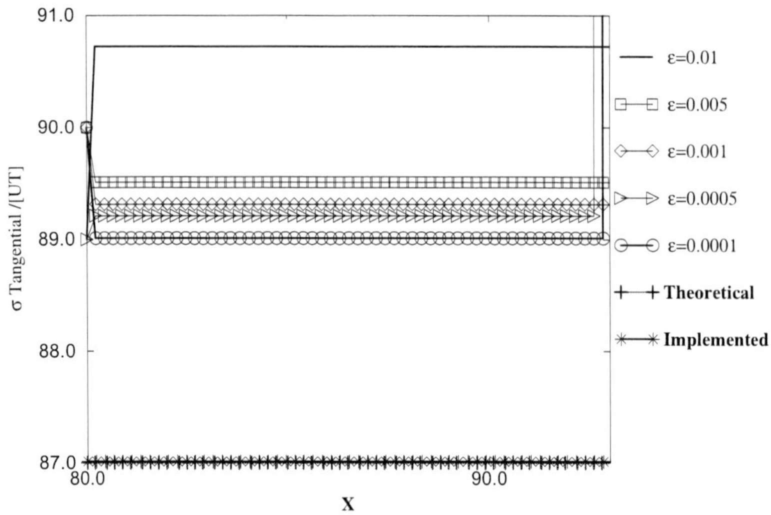

Figure 6. Ratio between tangential stress and tangential displacement (case d).

simplify the computations, the Lamé's coefficients of the joint are assumed to be given by

$$\begin{cases} \lambda_\varepsilon = \varepsilon^\gamma \lambda, & \gamma \geq 0, \\ \mu_\varepsilon = \varepsilon^\delta \mu, & 0 \leq \delta < 2. \end{cases}$$

The problem is solved with the following decreasing values of ε :

$$1.10^{-2},\ 5.10^{-3},\ 1.10^{-3},\ 5.10^{-4},\ 1.10^{-4}.$$

In order to analyze the results obtained, the displacement fields (u_N,u_T) and the stress vector (σ_N, σ_T) were computed on $S_\varepsilon^\pm$ in the case of the initial problem (and on S in that of the limit problem) in the nine cases studied. The theoretical and numerical curves of the stick and sliding nodes are then compared. The friction problem is solved by means of a fixed point algorithm coupled with a relaxation procedure [46]. Similar results were obtained (see fig. 5 and 6). As can be seen from these figures, the limit law can be taken to be valid for at thin layer thickness up to 10^{-3} mm. This value is suitable for a large class of problems (for example in glue-bonding processes).

5.3. Predictive Nature of the Model

In this section, we focus on a particular characteristic of the model, its predictive nature which depends on the mechanical and geometrical parameters of the glue and the adherents. The model presented in the previous paragraphs involves four parameters: the thickness of the glue ε, the stiffness coefficients of the glue λ and μ and the friction coefficient f.
We observe that the various cases depend both on the limits of λ/ε and of μ/ε. We stress

the fact that the nine laws that appear depend only on the relative values of the parameters. If mathematically, these limits are perfectly defined, mechanically, in a non academic problem, they are a priori unknown. The engineer only knows (or has the possibility of knowing) the thickness and the stiffness of the glue. The limits are approximated in a satisfactory way by the ratio between the stiffness and the thickness. In other words, the values of $\bar{\lambda}$ and $\bar{\mu}$ can be replaced by λ/ε and μ/ε respectively. But these approximations are never equal to zero or to infinity. So when shall we consider that these limit values are obtained ? Simply by comparison to a reference value. For example in the case of isotropic adherents, the reference values, would be λ_m/L or μ_m/L, where λ_m and μ_m are the moduli and L the characteristic length of the adherents (chosen equal to 1 in the numerical applications), with $\lambda << \lambda_m$, $\mu << \mu_m$ and $\varepsilon << L$, which expresses the fact that the interface is both soft and thin.
More precisely, when

- λ/ε and μ/ε are very small by comparison to the reference value, i.e. the glue is relatively more soft than thin, $\bar{\lambda}$ and $\bar{\mu}$ shall be taken equal to zero;
- λ/ε and μ/ε are very large by comparison to the reference value, i.e. the glue is relatively more soft than thin, $\bar{\lambda}$ and $\bar{\mu}$ shall be taken equal to infinity;
- λ/ε and μ/ε are finite by comparison to the reference value, i.e. the glue is as thin as soft, $\bar{\lambda}$ and $\bar{\mu}$ shall be taken equal to λ/ε and μ/ε respectively.

Numerically, we can observe that "very small" corresponds to less than 3% of the reference value and that "very large" corresponds to more than 30 times the reference value. The "finite case" corresponds to the range between these two extreme values.

6. Non Linear Imperfect Interface: Non Convex Energy

In this section, we consider:

- that the thin layer is governed by a non convex energy,
- that the stiffness of the glue is lower than that of the bodies,
- that the link between the glue and the bodies is perfect.

Non convex energies are usually associated to phase transforming materials [10, 19]

6.1. Constitutive Relation of a Phase Transforming Adhesive

To model the constitutive behavior of the adhesive, we adopt the Frémond constitutive relation, which involves three phases, the austenite and two variants of martensite [19]:

$$\left\{\begin{array}{lll} \sigma_{ij} & = & \lambda e_{kk}\delta_{ij} + 2\mu e_{ij} + \alpha\chi_{ij}, \ \text{if}\, \alpha\chi_{ij}e_{ij} \leq -c \\ \sigma_{ij} & = & \lambda e_{kk}\delta_{ij} + 2\mu e_{ij}, \ \text{if}\, |\alpha\chi_{ij}e_{ij}| \leq c \\ \sigma_{ij} & = & \lambda e_{kk}\delta_{ij} + 2\mu e_{ij} - \alpha\chi_{ij}, \ \text{if}\, \alpha\chi_{ij}e_{ij} \geq c, \end{array}\right. \quad in\ B^{\varepsilon} \tag{45}$$

and where constants α and c are two material parameters. The constant c can be related to a critical temperature, below which the stress-strain curve has only two ascending branches, each one corresponding to a different variant of martensite (fig. 7.1-a) Above the transformation temperature, a third phase exists, called austenite and corresponding to an intermediate ascending branch in the stress-strain curve (fig. 7.1-b).

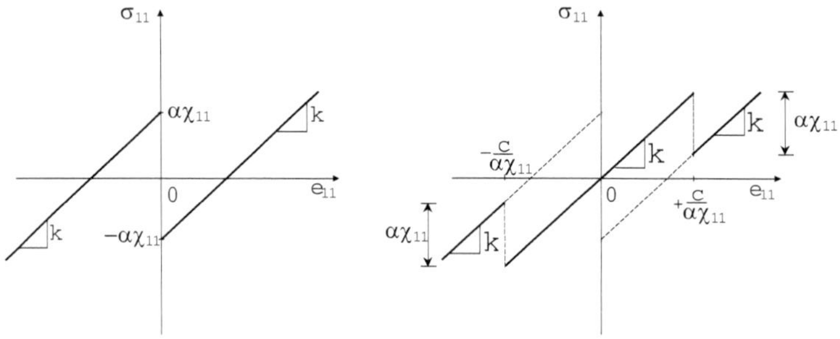

Figure 7. Stress-strain diagrams simulating an extension test on a phase transforming adhesive: (a) $c \leq 0$, (b) $c > 0$.

The problem is studied by means of matched asymptotic expansions. Let S, T denote the diagonal 2×2 matrices with $S_{11} = K_T$, $S_{22} = K_N$ and $T_{11} = T_{22} = K_{PT}$. Let x be the vector with $x_1 = \chi_{12}$ and $x_2 = \chi_{22}$. Note that the coefficients of S (resp. T) can be equal to infinity, zero or a non-zero bounded value (resp. infinity or a non-zero bounded value). Using these notations, we obtain the following family of contact laws

- $\bar{c} \leq 0$

$$\begin{aligned} \sigma n &= S[u] + Tx \quad if\ \alpha x_i[u_i] \leq 0, \\ \sigma n &= S[u] - Tx \quad if\ \alpha x_i[u_i] \geq 0, \end{aligned} \tag{46}$$

.

- $\bar{c} > 0$

$$\begin{aligned} \sigma n &= S[u] + Tx \quad if\ \alpha x_i[u_i] \leq -\bar{c}, \\ \sigma n &= S[u] \quad if\ |\alpha x_i[u_i]| \leq \bar{c}, \\ \sigma n &= S[u] - Tx \quad if\ \alpha x_i[u_i] \geq \bar{c}, \end{aligned} \tag{47}$$

.

6.2. Interface Laws Arising from Energy Minimization

In a one dimensional setting, it is possible to solve the equilibrium problem and discuss the stability of the equilibrium solutions for an adhesive obeying the constitutive relation

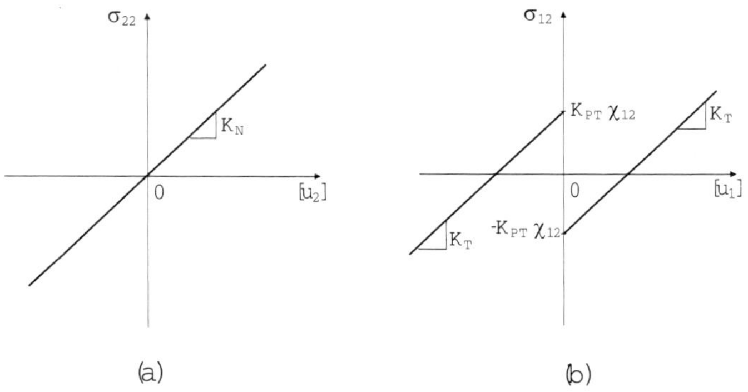

Figure 8. Contact laws for $c \leq 0$: (a) normal contact law for $\chi_{22} = 0$; (b) tangential contact law.

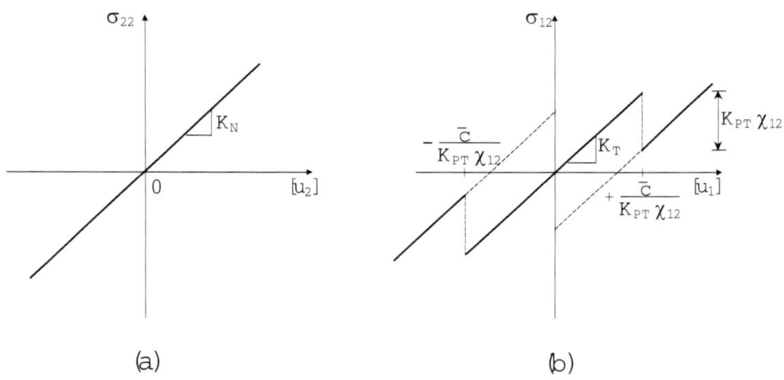

Figure 9. Contact laws for $c > 0$: (a) normal contact law for $\chi_{22} = 0$; (b) tangential contact law.

(45). This is done in the next subsection. The closed form solutions allow us to explicitly calculate the corresponding interface laws when the thickness of the adhesive tends to zero (see subsection 6.2.2).

6.2.1. Metastable States for a One Dimensional Phase Transforming Adhesive

Let us take a one-dimensional bar occupying a reference unstressed space configuration $\Omega = (0, l)$, starting from which there are displacement fields $u = u(x)$, $x \in (0, l)$ which are continuous, with the piecewise continuous derivative $u' \geq 0$. The bar is fixed at the extremity $x = 0$, and is subjected to a prescribed displacement $\delta > 0$ at $x = L$.

The bar is assumed to be made of two different elastic materials. In the range $0 \leq x \leq \varepsilon L$, the bar is composed of an adhesive layer characterized by the piecewise quadratic stored energy density

$$w_a(e) = \min_{e>0}\{\frac{k}{2}e^2; \frac{k}{2}e^2 - \alpha\chi e + c\} , \tag{48}$$

where $k > 0$ is the elasticity of the adhesive.

In the range $\varepsilon L \leq x \leq L$, the bar is composed of a material with quadratic stored energy density having an elastic modulus K

$$w_b(e) = \frac{K}{2}e^2 . \tag{49}$$

This system admits the following possible solutions:

i) If $0 \leq \sigma_\varepsilon < kc(\alpha\chi)^{-1} - \alpha\chi$, then,

$$\bar{u} = \begin{cases} \frac{\sigma_\varepsilon}{k}x & x \in [0, \varepsilon L] , \\ \frac{\sigma_\varepsilon}{K}x + \sigma_\varepsilon \varepsilon L(\frac{1}{k} - \frac{1}{K}) & x \in (\varepsilon L, L] , \end{cases} \tag{50}$$

with $\sigma_\varepsilon = \hat{k}_\varepsilon \delta L^{-1}$, and $\hat{k}_\varepsilon^{-1} := \varepsilon k^{-1} + (1-\varepsilon)K^{-1}$. Using the expression for σ_ε, it is easy to see that this solution is possible whenever $0 \leq \delta < (kc(\alpha\chi)^{-1} - \alpha\chi)L\hat{k}_\varepsilon^{-1}$. Since δ is positive, this solution exists only if $c > 0$ and corresponds to the case of adhesive consisting of only austenite.

ii) If $c > 0$ and $kc(\alpha\chi)^{-1} - \alpha\chi \leq \sigma_\varepsilon < kc(\alpha\chi)^{-1}$, then the adhesive is a mixture of austenite and martensite. Let $\lambda \in (0, 1)$ denote the austenite volume fraction. As λ varies in $[0, 1]$, we obtain a set of equilibrium solutions:

$$\bar{u} = \begin{cases} \frac{\sigma_\varepsilon}{k}x & x \in [0, \lambda\varepsilon L] , \\ \frac{\sigma_\varepsilon + \alpha\chi}{k}x - \frac{\alpha\chi\varepsilon\lambda L}{k} & x \in (\lambda\varepsilon L, \varepsilon L] , \\ \frac{\sigma_\varepsilon}{K}x + \varepsilon L(\frac{\sigma_\varepsilon + (1-\lambda)\alpha\chi}{k} - \frac{\sigma_\varepsilon}{K}) & x \in (\varepsilon L, L] , \end{cases} \tag{51}$$

with

$$\sigma_\varepsilon = \hat{k}_\varepsilon(\frac{\delta}{L} - \varepsilon(1-\lambda)\frac{\alpha\chi}{k}) . \tag{52}$$

Using this expression for σ_ε, it turns out that the two-phase solution exists whenever

$$((kc(\alpha\chi)^{-1} - \alpha\chi)\hat{k}_\varepsilon^{-1} + \varepsilon\alpha\chi(1-\lambda)k^{-1})L \leq \delta \leq (kc(\alpha\chi)^{-1}\hat{k}_\varepsilon^{-1} + \varepsilon\alpha\chi(1-\lambda)k^{-1})L . \tag{53}$$

Since the value of the volume fraction λ is within the $0 < \lambda < 1$ range, this condition can be further extended as follows

$$(kc(\alpha\chi)^{-1} - \alpha\chi)\hat{k}_\varepsilon^{-1}L \leq \delta \leq (kc(\alpha\chi)^{-1} + \varepsilon\alpha\chi k^{-1})L . \tag{54}$$

Therefore, given any $\lambda \in (0, 1)$, if the prescribed elongation δ lies in the above range, then there will exist in the adhesive a two-phase solution involving a mixture of both phases.

iii) If $\sigma_\varepsilon \geq kc(\alpha\chi)^{-1}$, then the solution is

$$\bar{u} = \begin{cases} \dfrac{\sigma_\varepsilon + \alpha\chi}{k} x & x \in [0, \varepsilon L] \,, \\ \dfrac{\sigma_\varepsilon}{K} x + \varepsilon L(\dfrac{\sigma_\varepsilon + \alpha\chi}{k} - \dfrac{\sigma_\varepsilon}{K}) & x \in (\varepsilon L, L] \,, \end{cases} \tag{55}$$

with $\sigma_\varepsilon = \hat{k}_\varepsilon(\delta L^{-1} - \alpha\chi\varepsilon k^{-1})$. This solution, which is possible for $\delta > kc(\alpha\chi)^{-1} L\hat{k}_\varepsilon^{-1} + \alpha\chi\varepsilon Lk^{-1}$ when $c > 0$ and for any positive δ when $c \leq 0$, describes the case of adhesive consisting entirely of martensite.

It is established that the configurations described by the solutions (50), (51) and (55) correspond to weak local minimizers [34]. It is also established that the following equilibrium configurations correspond to a global minimizer:

a) if $0 \leq \delta < (kc(\alpha\chi)^{-1} - \alpha\chi/2)L\, ke^{-1}$, then solution (50) is a global minimizer;

b) if $(kc(\alpha\chi)^{-1} - \alpha\chi/2)L\hat{k}_\varepsilon^{-1} \leq \delta < (kc(\alpha\chi)^{-1} - \alpha\chi/2)L\hat{k}_\varepsilon^{-1} + \varepsilon\alpha\chi Lk^{-1}$, then the solution (51) with σ_ε given by the Maxwell stress $\sigma_M = kc(\alpha\chi)^{-1} - \alpha\chi/2$ is a global minimizer. Substituting σ_M into (52), we find the austenite volume fraction determined solely for the given δ :

$$\lambda_{glo} = 1 + \frac{K}{\varepsilon\alpha\chi}\left(\frac{k}{\hat{k}_\varepsilon} c(\alpha\chi)^{-1} - \frac{\alpha\chi}{2\hat{k}_\varepsilon} - \frac{\delta}{L}\right) . \tag{56}$$

Note that when $\delta = (kc(\alpha\chi)^{-1} - \alpha\chi/2)L\hat{k}_\varepsilon^{-1}$ we have $\lambda_{glo} = 1$, i.e., the adhesive is still in the austenite phase. If δ is continuously increased, the austenite volume fraction decreases and the stress in the bar remains constant and equal to the Maxwell stress. When $\delta = (kc(\alpha\chi)^{-1} - \alpha\chi/2)L\hat{k}_\varepsilon^{-1} + \varepsilon\alpha\chi Lk^{-1}$, we have $\lambda_{glo} = 0$, and the adhesive has completed the transformation from austenite to martensite.

c) if $\delta \geq (kc(\alpha\chi)^{-1} - \alpha\chi/2)L\hat{k}_\varepsilon^{-1} + \varepsilon\alpha\chi Lk^{-1}$, then the solution (55) is a global minimizer.

6.2.2. Interface Laws for a Thin Phase Transforming Adhesive

To obtain the interface law we study the asymptotic behavior of the above equilibrium solutions when both the parameters (ϵ, k) tend to zero and the thin adhesive layer is replaced by a point. To study the asymptotic behavior of the adhesive, we set

$$k = \bar{k}\, \varepsilon \,. \tag{57}$$

Our aim here is to study the relations between the limits

$$\sigma = \lim_{\varepsilon\to 0} \sigma_\varepsilon \,, \quad [u] = \lim_{\varepsilon\to 0} (\bar{u}(\varepsilon L) - \bar{u}(0)) \,, \tag{58}$$

where σ_ε and $\bar{u}$ correspond to the equilibrium configurations determined in the previous subsection. This study leads to defining an asymptotic contact law linking the limit stress in the bar, σ_ε, to the jump in the displacement occuring at the adhesive interface, $[u]$. This contact law describes the limit behavior of the adhesive.

In addition to (57), we need to specify the scaling of the material parameters with ε. To make the limits (58) finite, we take

$$c = \bar{c}\,\varepsilon^{-1}, \quad \alpha = \bar{\alpha}\,, \quad \chi = \bar{\chi}\,, \tag{59}$$

where $\bar{k}$, $\bar{c}$, $\bar{\alpha}$ and $\bar{\chi}$ are independent of ε.

Substituting (57), (59) into the expressions for σ_ε and $\bar{u}$ listed in section 2 in (*a*), (*b*) and (*c*), taking the limit $\varepsilon \to 0^+$ and eliminating δ between σ_ε and $[u]$, we obtain the following contact law:

$$\sigma_{glo} = \begin{cases} \bar{k}\dfrac{[u]}{L}, & 0 \le [u] < L(\bar{c}(\bar{\alpha}\bar{\chi})^{-1} - \dfrac{\bar{\alpha}\bar{\chi}}{2\bar{k}})\,, \\ (\bar{k}\bar{c}(\bar{\alpha}\bar{\chi})^{-1} - \dfrac{\bar{\alpha}\bar{\chi}}{2}), & L(\bar{k}\bar{c}(\bar{\alpha}\bar{\chi})^{-1} - \dfrac{\bar{\alpha}\bar{\chi}}{2\bar{k}}) \le [u] < L(\bar{k}\bar{c}(\bar{\alpha}\bar{\chi})^{-1} + \dfrac{\bar{\alpha}\bar{\chi}}{2\bar{k}})\,, \\ \bar{k}\dfrac{[u]}{L} - \bar{\alpha}\bar{\chi}, & [u] \ge L(\bar{k}\bar{c}(\bar{\alpha}\bar{\chi})^{-1} + \dfrac{\bar{\alpha}\bar{\chi}}{2\bar{k}})\,, \end{cases} \tag{60}$$

corresponding to global minimizers of the original equilibrium problem. In the same way, taking σ_ε and $\bar{u}$ as in (*i*), (*ii*) and (*iii*), we can calculate the limit contact law corresponding to local minimizers. Note that this law turns out to be undefined, because of the lack of information available due to the non uniqueness of the local minimizers. Indeed, we obtain

$$\sigma_{loc} = \begin{cases} \bar{k}\dfrac{[u]}{L}, & 0 \le [u] < L(\bar{c}(\bar{\alpha}\bar{\chi})^{-1} - \dfrac{\bar{\alpha}\bar{\chi}}{2\bar{k}})\,, \\ \bar{\Sigma}, & L(\bar{k}\bar{c}(\bar{\alpha}\bar{\chi})^{-1} - \dfrac{\bar{\alpha}\bar{\chi}}{2\bar{k}}) \le [u] < L(\bar{k}\bar{c}(\bar{\alpha}\bar{\chi})^{-1} + \dfrac{\bar{\alpha}\bar{\chi}}{2\bar{k}})\,, \\ \bar{k}\dfrac{[u]}{L} - \bar{\alpha}, & [u] \ge L(\bar{k}\bar{c}(\bar{\alpha}\bar{\chi})^{-1} + \dfrac{\bar{\alpha}\bar{\chi}}{2\bar{k}})\,, \end{cases} \tag{61}$$

where $\bar{\Sigma}$ can take any value in $[\bar{k}\bar{c}(\bar{\alpha}\bar{\chi})^{-1} - \bar{\alpha}\bar{\chi},\ \bar{k}\bar{c}(\bar{\alpha}\bar{\chi})^{-1} + \bar{\alpha}\bar{\chi}]$. Therefore, local minimizers give rise to multiple contact laws, all of which are included in the dashed parallelogram depicted in Figure 10. Note the non uniqueness of the equilibrium solutions, due to the non convexity of the deformation energy associated to (45). To rule out non uniqueness, one can postulate a nucleation condition and a kinetic relation, specifying the path inside the parallelogram of the possible states [2].

7. An Example of Numerical Implementation

This section is devoted to the numerical solution of the problem presented in the previous section. Note that there exist several studies to solve this kind of problems [3, 5–7, 12, 13, 28–30, 32, 46]. In what follows, we will adopt the two-dimensional context, focusing on the contact between a deformable solid A occupying Ω and a rigid body. In this case,

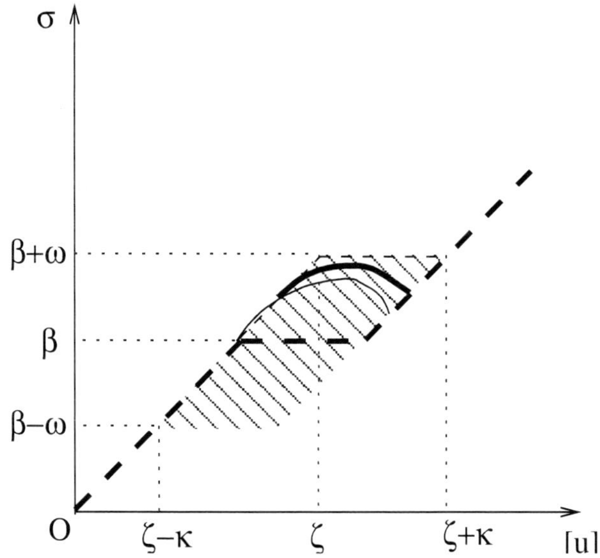

Figure 10. Contact law obtained by energy minimization. To make the figure clearer, we have defined $\zeta = c(\bar{\alpha}\bar{\chi})^{-1}$, $\kappa = \bar{\alpha}\bar{\chi}\bar{k}^{-1}$, $\beta = \bar{k}(\bar{\alpha}\bar{\chi})^{-1} - \bar{\alpha}\bar{\chi}/2$, $\omega = \bar{\alpha}\bar{\chi}/2$. The shaded region is the contact law domain corresponding to local minimizers. Contact law corresponding to global minimizers (Maxwell line): - -. Contact laws corresponding to local minimizers: — linear kinetic, — pinning kinetic, - - convex decomposition.

Problem 2

$$(P)\begin{cases} Find\ (u,\sigma)\ such\ that: & \\ \sigma_{ij,j} = -\varphi_i & in\ \Omega \\ \sigma_{ij} = a_{ijkh}e_{kh}(u) & in\ \Omega \\ u = 0 & on\ \Gamma_0 \\ \sigma n = g & on\ \Gamma_1 \\ \sigma_N \leq 0 & on\ S \\ \sigma_N = -K_N u_N^+ & on\ S \\ |\sigma_T| \leq f|\sigma_N| & on\ S \\ \quad If\ |\sigma_T| < f|\sigma_N|\ then\ \sigma_T = -K_T u_T & \\ \quad If\ |\sigma_T| = f|\sigma_N|\ then\ u_T = -\zeta\sigma_T & with\ \zeta \geq 0 \end{cases}$$

The contact laws are shown in figs. 11 and 12. In lines with previous studies, the above problem is equivalent to:

Problem 3

Find ρ, the fixed point of the application $\rho \longrightarrow f|\sigma_N(u(\rho))|$, where $u = u(\rho)$ is the solution of

$$\begin{cases} Find\ u \in V\ such\ that: & \\ J(u) \leq J(v), & \forall v \in V, \end{cases}$$

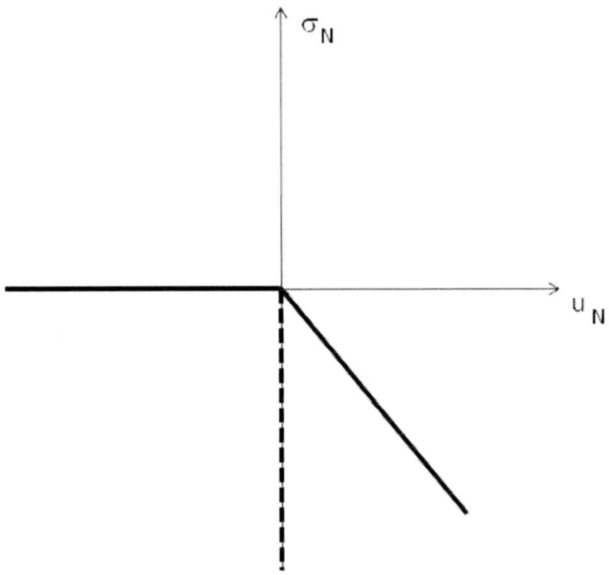

Figure 11. Contact law: normal component (the dotted line corresponds to an infinite value of K_N).

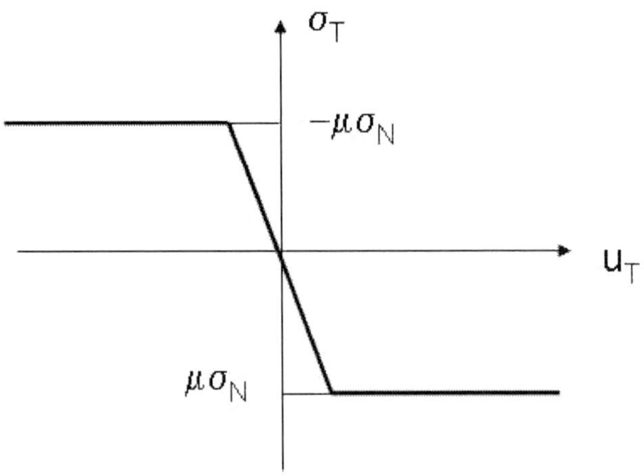

Figure 12. Contact law: tangential component (the dotted line corresponds to an infinite value of K_T).

with

$$J(v) = \frac{1}{2}\int_\Omega ae(v)e(v)\,dx - \int_\Omega \varphi.u\,dx + \int_{\Gamma_1} g.u\,dl + \int_S \phi(v_N)\,ds + \int_S \psi(v_T)\,ds,$$

where ϕ is given by

$$\phi(v_N) = \frac{K_N}{2}(v_N^+)^2,$$

and ψ is given by

$$\psi(v_T) = \frac{1}{2}K_T.(v_T)^2 + \rho(|v_T| - \frac{\rho}{K_T})^+.$$

Problem 3 is discretized using a finite element method formulated in terms of the displacements. We usually adopt $P1$ finite elements (triangles with three nodes and six degrees of freedom) or $Q1$ finite elements (quadrangles with four nodes and eight degrees of freedom). We have to minimize a functional still denoted J for v in R^{2NP} such that

$$\mathcal{J}(v) = \frac{1}{2}v^T Av + \frac{1}{2}(Pv)^T B(Pv) - v^T L.$$

The following notations are used:
NP : total number of nodes,
NC : number of contact nodes,
I_N: indices of normal components of contact nodes,
I_T: indices of tangential components of contact nodes,
A: stiffness matrix associated with volume terms with coefficient a_{ij},
B: stiffness matrix to the surface terms with coefficient b_{ij},
P: projection from R^{2NP} to R^{2NC},
L: generalized loading vector with coefficients L_i.

Relaxation method with constraints The relaxation method consists in finding the solution to problem 3 by solving a sequence of minimization problems in R^{2NP}

$$\begin{cases} Find\ u_i^{n+\frac{1}{2}}\ such\ that\ \forall v \in R^{2NP} \\ \mathcal{J}(u_1^{n+1}, \ldots, u_{i-1}^{n+1}, u_i^{n+\frac{1}{2}}, u_{i+1}^n, \ldots, u_{2NP}^n) \leq \mathcal{J}(u_1^{n+1}, \ldots, u_{i-1}^{n+1}, v, u_{i+1}^n, \ldots, u_{2NP}^n). \end{cases}$$

ω is taken to denote the relaxation coefficient.

First we deal with the normal components. When $i \in I_N$, the algorithm is written in the following form:

$$\left\{\begin{array}{l} u_i^{n+\frac{1}{2}} = \dfrac{1}{d_{ii}^{n+\frac{1}{2}}}(L_i - \sum_{j=1}^{i-1} d_{ij}^{n+1} u_j^{n+1} - \sum_{j=i+1}^{2NP} d_{ij}^{n} u_j^{n}) \\ with \\ d_{ij}^n = \left\{\begin{array}{ll} a_{ij} + \gamma(u_j^n) b_{ij} & if\ j \in I_N \\ a_{ij} + \eta(u_j^n) b_{ij} & if\ j \in I_T \\ a_{ij} & otherwise \end{array}\right. , \\ \gamma(u) = \left\{\begin{array}{l} 0 \quad if\ u \leq 0 \\ 1 \quad if\ u > 0 \end{array}\right. , \\ and\ \eta(u) = \left\{\begin{array}{l} 0 \quad if\ |u| > \dfrac{\rho}{K_T} \\ 1 \quad if\ |u| \leq \dfrac{\rho}{K_T} \end{array}\right. , \\ u_i^{n+1} = (1-\omega) u_i^n + \omega u_i^{n+\frac{1}{2}} \end{array}\right. .$$

As regards the tangential components, when $i \in I_T$ we first take the fixed point problem $\rho_i^{l+1} = f|\sigma_N(u(\rho_i^l))|$ and then write

$$\left\{\begin{array}{l} u_i^{n+\frac{1}{2}} = \dfrac{1}{d_{ii}^{n+\frac{1}{2}}}(L_i - \sum_{j=1}^{i-1} d_{ij}^{n+1} u_j^{n+1} - \sum_{j=i+1}^{2NP} d_{ij}^{n} u_j^{n} - \theta(u_i^{n+\frac{1}{2}}).\rho_i^l) \\ with \\ \theta(u) = \left\{\begin{array}{ll} -1 & if\ u < \dfrac{-\rho}{K_T} \\ 1 & if\ u > \dfrac{\rho}{K_T} \\ 0 & if\ |u| \leq \dfrac{\rho}{K_T} \end{array}\right. , \\ u_i^{n+1} = (1-\omega) u_i^n + \omega u_i^{n+\frac{1}{2}} \end{array}\right. .$$

Testing the validity of the algorithm: compression of a bar In this paragraph the algorithm is tested and its validity is confirmed. We used a benchmark test developed by the group working on "Validation of computer codes" at the French Research Group "Large Deformations and Damage" [46].

Here we adopt the context of plane strains and take the case of a long bar with a square section (fig. 13) and Lamé's coefficients $\lambda = 45GPa$ and $\mu = 54GPa$. The contact zone (interface law) corresponds to the part AD with a friction coefficient equal to $f = 1$. $u_1 = 0$ on DE and $u_1 = u_2 = 0$ at point D. The loading $F1 = 10\ daN/mm^2$ is imposed on AG and $F2 = -5\ daN/mm^2$ on GE.

Using the finite element method, the contact zone is discretized by 32 nodes. The changes in contact status are given in table 1 at different values of $\overline{\mu}$ and $\overline{\lambda}$. If we compare the re-

Table 1. Behavior of the interface for some stiffness

Stiffness	Gap AB	Sliding $\lvert\sigma_T\rvert = f\lvert\sigma_N\rvert$ BC	Stick $\lvert\sigma_T\rvert < f\lvert\sigma_N\rvert$ CD
(1) $\overline{\mu} = 5.410^{+8}$; $\overline{\lambda} = 4.510^{+8}$	3 nodes	16 nodes	13 nodes
(2) $\overline{\mu} = 5.410^{+2}$; $\overline{\lambda} = 4.510^{+8}$	3 nodes	14 nodes	15 nodes
(3) $\overline{\mu} = 5.410^{+2}$; $\overline{\lambda} = 4.510^{+2}$	3 nodes	13 nodes	16 nodes
(4) $\overline{\mu} = 5.410^{-2}$; $\overline{\lambda} = 4.510^{+8}$	0 nodes	00 nodes	32 nodes
(5) $\overline{\mu} = 5.410^{-2}$; $\overline{\lambda} = 4.510^{+2}$	0 nodes	00 nodes	32 nodes
(6) $\overline{\mu} = 5.410^{-2}$; $\overline{\lambda} = 4.510^{-2}$	0 nodes	00 nodes	32 nodes

sults obtained with those published in [46] (Signorini-Coulomb laws), it can be seen that at $\overline{\lambda} = 4.510^{+8}$ and $\overline{\mu} = 5.410^{+8}$ the present results are similar to those obtained by [46]. These coefficients, which are very large (i.e. "equal" to infinity), correspond to the limit case (Signorini-Coulomb). The decrease in the values of $\overline{\mu}$ corresponds to the increase in the stick zone (13, 15 and 32 nodes). Note that for this example and for cases presented in figure 14, the normal displacements are slightly changed by the variations of K_N and K_T. In particular, the penetration is negligible. However, as observed on figure 15, for this problem the coefficients variations strongly affect the tangential displacements. A low value of the coefficient K_T reflects in an increase of the tangential displacement. The nodes slide without reaching the value of Coulomb's sliding limit (fig. 15). Note that numerical tests have shown the robustness of this algorithm [49].

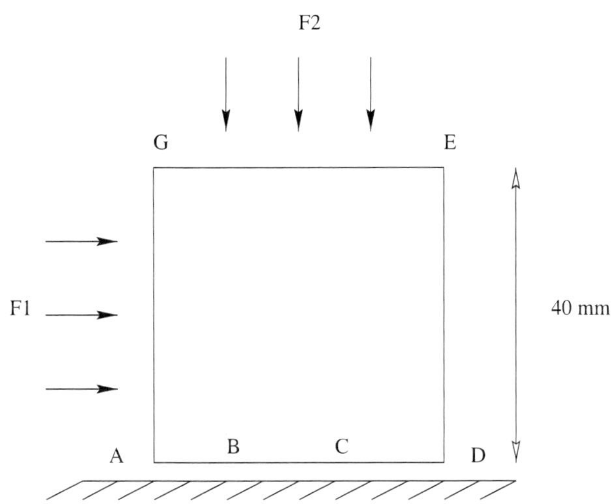

Figure 13. The problem of the long bar.

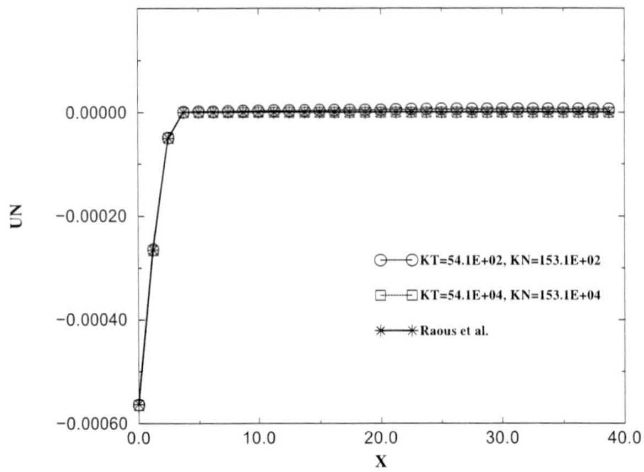

Figure 14. Normal displacement with respect to X.

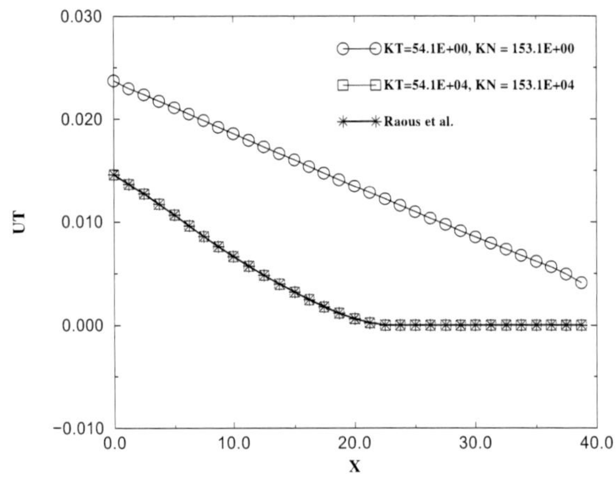

Figure 15. Tangential displacement with respect to X.

8. Conclusion

In this chapter, we analyze the asymptotic behavior of a thin adhesive. To obtain the interface law, which describes the behavior of an adhesive of vanishing thickness, different approaches have been tested. First, we have dealt with adherents and adhesive having similar rigidity and performed a Γ-convergence analysis to obtain the relation between the stress vector and the displacement fields. Next, we have used matched asymptotic expansions to analyze the asymptotic behavior of a a two-dimensional equilibrium problem including frictional contact at the interface between the thin adhesive and the adherents. Lastly, we have studied the problem of a thin layer with a non convex energy, which is used to model phase transformation of the adhesive. To conclude, we have presented a numerical algorithm able to solve the problem with frictional contact.

The asymptotic methods presented in the chapter allow to obtain a precise description of the limit behavior of a vanishing adhesive in terms of an interface law. The cases presented show that different constitutive behaviors of the adhesive give raise to different interface laws. The mathematical methods presented in the chapter provide a precise limit behavior if the constitutive relation of the adhesive is known, an occurrence which is rarely satisfied in practice. Another delicate point is that the methods illustrated in the chapter can be used under the assumption that the model of continuum is valid at the scale of the adhesive. There are cases in which the adhesive behavior is better described by means of molecular dynamics and other techniques should then be applied [43, 51].

References

[1] R. Abdelmoula, M. Coutris, J. Marigo, "Comportement asymptotique d'une interphase élastique mince", *Compte Rendu Académie des Sciences Série II*, **326**, 237-242, 1998.

[2] R. Abeyaratne, K. Bhattacharya, J. K. Knowles, "Strain-energy functions with multiple local minima:modeling phase transformations using finite thermoelasticity", in *Nonlinear elasticity: Theory and applications* (ed. Y. Fu and R.W. Ogden), Cambridge University Press, 433-490, 2001.

[3] K. Ach, P. Alart, M. Barboteu, F. Lebon, B. Mbodji, "Parallel frictional contact algorithms and industrial applications", *Computer Methods in Applied Mechanics and Engineering,* **177**,169-181, 1999.

[4] A. Ait-Moussa, "Modélisation et étude des singularités d'un joint collé", PhD thesis, Université Montpellier II, 1989.

[5] P. Alart, M. Barboteu, F. Lebon, "Solution of frictional contact problems by an EBE preconditioner", *Computational Mechanics,* **20**, 370-378, 1997.

[6] P. Alart, A. Curnier, "A mixed formulation for frictional contact problems prone to Newton like solution methods", *Computer Methods in Applied Mechanics and Engineering,* **92**, 353-375, 1991.

[7] P. Alart, F. Lebon, "Solution of frictional contact problems using ILU and coarse/fine preconditioners", *Computational Mechanics,* **16**, 98-105, 1995.

[8] G. Amontons, "Sur l'origine de la résistance dans les machines", *Mémoire de l'Académie Royale,* 206-222, 1699.

[9] H. Attouch, "Variational Convergence for Functions and Operators", Pitman Advanced Publishing Program, 1984.

[10] J. Ball, R. D. James,"Fine phase mixtures as minimizers of energy", *Archive for Rational Mechanics and Analysis,* **100**, 13-52, 1987.

[11] Y. Benveniste, T. Miloh, "Imperfect soft and stiff interfaces in two-dimensional elasticity", *Mechanics of Materials,* **33**, 309-323, 2001.

[12] P. Bisegna, F. Lebon, F. Maceri, "D-PANA: a convergent block-relaxation solution method for the discretized dual formulation of the Signorini-Coulomb contact problem", *Comptes Rendus Académie des Sciences, Série I*, **333**, 1053-1058, 2001.

[13] P. Bisegna, F. Lebon, F. Maceri, "Relaxation procedures for solving Signorini-Coulomb contact problems", *Advances in Engineering Software,* **35**, 595-600, 2004.

[14] D. Caillerie, "The effect of a thin inclusion of high rigidity in an elastic body", *Mathematical Methods in Applied Sciences,* **2**, 251-270, 1980.

[15] C.A. Coulomb, "Théorie de machines simples", *Mémoire de Mathématiques et Physique de l'Académie Royale,* **10**, 161-342, 1785.

[16] G. Dal Maso, "An Introduction to Gamma-Convergence", Birkhauser, 1993.

[17] G. Del Piero, R. Rizzoni, "Weak local minimizers in finite elasticity",*Journal of Elasticity,* **93** (3), 203-244, 2008.

[18] W. Eckhaus, "Asymptotic analysis of singular perturbations", North-Holland, Amsterdam, 1979.

[19] M. Frémond, "Matériaux à mémoire de forme", *Compte Rendu Académie des Sciences,* **304**, 239-244, 1987.

[20] J.F. Ganghoffer, A. Brillard, J. Schultz, "Modelling of the mechanical behaviour of joints bonded by a nonlinear incompressible elastic adhesive", *European Journal of Mechanics A/Solids,* **16**, 255-276, 1997.

[21] G. Geymonat and F. Krasucki, "Analyse asymptotique du comportement en flexion de deux plaques collées", *Compte Rendu Académie des Sciences Série* I, 325, 307-314, 1997.

[22] G. Geymonat, F. Krasucki, S. Lenci, "Mathematical analysis of a bonded joint with a soft thin adhesive", *Mathematics and Mechanics of Solids,* **16**, 201-225, 1999.

[23] Z. Hashin, "Thermoelastic properties of particulate composites with imperfect interfaces", *Journal of Mechanics and Physics of Solids,* **39**, 745-762, 1992.

[24] Z. Hashin, "Extremum principles for elastic heterogeneous media with imperfect interfaces and their application to bounding of effective moduli", *Journal of Mechanics and Physics of Solids* **40**, 767-781, 1992.

[25] Z. Hashin, "Thin interphase/imperfect interface in elasticity with application to coated fiber composites", *Journal of Mechanics and Physics of Solids* **50**, 473-486, 2002.

[26] A. Klarbring, "Derivation of the adhesively bonded joints by the asymptotic expansion method", *International Journal of Engineering Science,* **29**, 493-512, 1991.

[27] F. Krasucki, A. Munch, and Y. Ousset, "Analyse asymptotique d'un assemblage colle en elasticite non linéaire", *Compte Rendu Académie des Sciences Série IIb*, **329**, 429-434, 2001.

[28] F. Kuss, F. Lebon, "Méthodes duales pour le contact frottant", European *Journal of Computational Mechanics*, **16**, 33-51, 2007.

[29] F. Kuss, F. Lebon, "Comparison of numerical procedures for dual contact problems", *Advances in Engineering Software*, **40**, 697-706, 2009.

[30] F. Lebon, "Contact problems with friction: Models and simulations", *Simulation, Modelling, Practice and Theory,* **11**, 449-464, 2003.

[31] F. Lebon, A. Ould-Khaoua, and C. Licht, "Numerical study of soft adhesively bonded joints in finite elasticity", *Computational Mechanics,* **21**, 134-140, 1997.

[32] F. Lebon and M. Raous, "Friction modelling of a bolted junction under internal pressure loading", *Computers and Structures,* **43**, 925-933, 1992.

[33] F. Lebon, R. Rizzoni, S. Ronel-Idrissi, "Analysis of non-linear soft thin interfaces, *Computers and Structures*", **82**, 1929-1938, 2004.

[34] F. Lebon, R. Rizzoni, "Asymptotic study of soft thin layer: the non convex case", *Mechanics of Advanced Materials and Structures,* **15**, 12-20, 2008.

[35] F. Lebon, R. Rizzoni, "Asymptotic analysis of a thin interface: the case involving similar rigidity", *International Journal of Engineering Sciences,* **48**, 473486, 2010.

[36] F. Lebon, R. Rizzoni, "Asymptotic behavior of a hard thin linear elastic interphase: an energy approach", *International Journal of Solids and Structures* , **48**, 441-443, 2011.

[37] F. Lebon, S. Ronel-Idrissi, "Asymptotic studies of Mohr-Coulomb and Drucker-Prager soft thin layers", *International Journal of Steel and Composite Structures* , **4**, 133-148, 2004.

[38] F. Lebon, S. Ronel-Idrissi, "First order numerical analysis of linear thin layers", *ASME Journal of Applied Mechanics,* **74**, 824-828, 2007.

[39] F. Lebon, F. Zaittouni, "Asymptotic modelling of interface taking into account contact conditions: Asymptotic expansions and numerical implementation", *International Journal of Engineering Sciences,* **48**, 111-127, 2010.

[40] C. Licht, "Comportement asymptotique d'une bande dissipative mince de faible rigidité", *Compte Rendu Académie des Sciences Série I,* **317**, 429-433, 1993.

[41] C. Licht and G. Michaille. Une modélisation du comportement d'un joint collé élastique", *Compte Rendu Académie des Sciences Série I,* **322**, 295-300, 1996.

[42] C. Licht and G. Michaille, "A modeling of elastic adhesive bonded joints", *Advances in Mathematical Sciences and Applications,* **7**, 711-740, 1997.

[43] B. Q. Luan, S. Hyun, J.F. Molinari, N. Berstein, M. O. Robbins, "Multiscale modeling of two-dimensional contacts", *Physical Review E,* **74** (4, Part 2), 1-11, 2006.

[44] A. Ould-Khaoua, F. Lebon, C. Licht, and G. Michaille, "Thin layers in elasticity: a theoretical and numerical study", In ASME, editor, *Proceedings of the 1996 ESDA Conference*, **4**, 171-178, 1996.

[45] C. Pelissou, F. Lebon, "Asymptotic modeling of quasi-brittle interfaces", *Computers and Structures,* **87**, 1216-1223, 2009.

[46] M. Raous, P. Chabrand, and F. Lebon, "Numerical methods for solving unilateral contact problem with friction", *Journal of Theoretical and Applied Mechanics,* **7**,, 111-128, 1988.

[47] J. Sanchez-Hubert and E. Sanchez-Palencia, "Introduction aux méthodes asymptotiques et à l'homogénisation", Masson, 1992.

[48] P. Suquet, "Discontinuities and plasticity", In Springer-Verlag, editor, *Non- smooth mechanics and applications*, 315-323, 1988.

[49] F. Zaittouni, "Modélisation théorique et numérique d'interfaces", PhD thesis, Université Montpellier II, 2000.

[50] F. Zaittouni, F. Lebon, and C. Licht, "Etude théorique et numérique du comportement d'un assemblage de plaques", *Compte Rendu Académie des Sciences Série Mécanique,* **330**, 359-364, 2002.

[51] S. P. Xiao, T. Belytschko, "A bridging domain method for coupling continua with molecular dynamics", *Computer Methods in Applied Mechanics and Engineering* , **193** (17-20), 1645-1699, 2004.

In: Adhesive Properties in Nanomaterials, Composites ... ISBN: 978-1-61209-268-3
Editors: K. A. Wilkinson and D. A. Ordonez

Chapter 7

DURABILITY OF ADHESIVES AND MATRICES FOR POLYMER COMPOSITES USED IN RESTORATION AND REHABILITATION OF BUILDING STRUCTURES UNDER NATURAL AND ACCELERATED WEATHERING CONDITIONS†

Mariaenrica Frigione*
Department of Engineering for Innovation,
University of Salento, 73100 Lecce, Italy

ABSTRACT

The success of the fiber-reinforced polymer (FRP) systems in the restoration and rehabilitation of civil and monumental structures is due to their excellent properties, generally superior than those of traditional building materials. Of great importance, however, is the behavior of the repaired structure under loading and its durability in the outside climate. The lack of specific standards for durability investigation of materials employed in such applications makes difficult the assessment of reliable theoretical models. As an example, the available standard tests for adhesives generally refer to resins cured at elevated temperatures, neglecting the peculiarities of "cold-cured" adhesives.

In this chapter, the durability of the base components of FRP specifically designed for civil engineering industry, is reviewed. The most common environmental agents, mostly responsible for the deterioration of the materials performance, are examined. Finally, standardized accelerating tests are discussed as an effective method to predict the long term behavior of the weathered materials.

Keywords: Cold-cured thermosetting resins; Durability.

---

† A version of this chapter also appears in *Encyclopedia of Polymer Composites: Properties, Performance and Applications*, published by Nova Science Publishers, Inc. It was submitted for appropriate modifications in an effort to encourage wider dissemination of research.

* Corresponding author: mariaenrica.frigione@unisalento.it

INTRODUCTION

Fiber reinforced polymers (FRP) are currently used to restore and renew the built infrastructures, ranging from use as reinforcing elements in concrete and externally bonded elements for the strengthening of deteriorated and under-strength concrete and steel components, to use as structural components in building frames and bridge decks. As indicated by the JSCE research committee, several studies have been performed since 1980's on the utilization of FRP as concrete reinforcement, mainly in Japan, where FRP have been largely applied to existing structures, North America and Europe. [JSCE, 1993]

As a recent example, the use of composites based on polymeric thermosetting resins to increase the bearing capacity of damaged structures is becoming more and more widespread. The retrofit technique of concrete structures by means of FRP composites is addressed to improve the flexural and/or shear strength and stiffness of beams and to provide confinement to concrete elements (columns) under compression actions. [Norris et al., 1997; Mirmiran and Shahawy, 1997] Usually, FRP composites are externally bonded to concrete elements with polymeric thermosetting adhesives.

Due to a variety of different properties, FRP materials are being preferred to traditional construction materials, such as steel and concrete. The attractiveness of composites is not only based on performance attributes, such as high specific strength and specific stiffness, higher durability against corrosion, but also on their lower weights, ease of installation and reduced manufacture time. Wider choice of materials are available and product anisotropy can be tailored to specific application. There are, however, aspects of this technology that need further research and development, particularly the improvement in fire resistance, reduction in manufacturing cost, standardization of procedures and products, and a reliable assessment of their service life. The most pressing problem to be solved concerns the lack of adequate knowledge about the long term behavior of these materials, in particular when exposed to severe environment conditions [Pilakoutas et al., 1994; Karbhari, 2002]. Existing data on the durability performances of these materials in this specific field are scarce, not well documented and often contradictory. This issue hampers the enormous potential of composites in structural building construction applications, since the acceptable lifetime of products employed in this field should be in the order of 50-100 years. [Uomoto, 2003; Karbhari, 2005] There is, therefore, a need to find reasonable tools not only for the prediction of property changes of these materials with time but also for the determination of the remaining service-life of a structure working in widely variable service conditions.

As an example, the resin matrix can allow the ingress of moisture and this can lead to a variety of events, leading to the deterioration of the polymer and, in some cases, of the reinforcing fiber. Although carbon fibers are generally considered to be inert to most environmental influences likely to be faced in civil infrastructure applications, the inertness does not apply to the fiber-matrix bond. Glass fibers are well known to undergo stress corrosion in presence of moisture. [Mascia and Allavena, 1976]

Analyzing the behavior of polymeric composites under service conditions, the properties of any single element (primer, adhesive, fibers) have been analyzed in some depth only in recent times and the informations about the loss of characteristics due to external agents are in most cases still inadequate.

In this chapter, the different environmental agents, commonly encountered in the service life of these systems, are examined. The behavior of single components of FRP structures and adhesives, specifically designed for civil engineering industry, is reviewed. Moreover, different standardized accelerating tests are presented, in order to assess if they can be regarded as a meaningfully and effective method to predict the long term behavior of these materials when subjected to different environmental conditions.

ENVIRONMENTAL AGENTS

The performance and the long term behavior of thermoset polymer composites depends on the chemical nature of the base materials, on the conditions in which the matrix resin sets and hardens, on the process used to manufacture and to apply the composite and, mostly, on the environmental conditions in service. An environment can be defined as the sum of all factors acting on the material (such as: temperature cyclical changes, presence of water and of chemicals, prolonged freeze, freeze-thaw, alkaline environment, ultraviolet radiations, pollution, etc.). For indoor applications, the influence of humidity, changes in temperature and their combinations on the properties of composites must be taken into account. For the more common outdoor applications, on the other hand, it can be hypothesized that, during its service life, a polymeric composite will come into contact with atmospheric humidity, solar radiation, variations in temperature, acid rain, sea-water, deicing chemicals and alkaline environment when in the proximity of Portland cement concrete. Finally, polymer composites can be accidentally exposed to extreme environments, such as: fire, earthquake, explosive blasts. The influences of the latter extraordinary conditions on the performance of composites, however, will not be considered in this review.

Many researchers advise that the strength of the adhesive equilibrated with the "worst-case" environment is the key to achieve an effective design. [Renton, 1978; Romanko and Knauss, 1980; Albrecht et al., 1985] This is normally implemented in laboratory tests conducted at high temperatures on specimens pre-equilibrated with high levels of water vapor or liquid water. The specific conditions, however, must be carefully selected. As an example, for the application of adhesives to steel bridges in the USA, Albrecht *et al* [Albrecht et al., 1985] selected a test environment of 49°C and 90% R.H. .

The environmental conditions that the materials can encounter depend on several parameters that are not always predictable, such as the latitude, the season, the distance from sea, the local weather. [Wypych, 1995] Close to the equator, temperatures do not vary much and the radiative flux is uniform. Large temperature variations, on the other hand, occur in a continental interior, i.e. location far from coast. The surface temperature of a material is greatly affected by the absorption of radial energy. It can easily exceed ambient temperature by as much as 30°C. Wind speed can also have considerable effect on surface temperature, increasing the rate of heat exchange. Wind can also affect moisture concentration, dust deposition and can generally influence the degradation rate. Humidity and the amount of precipitation are more local than global climatic phenomena. Among the global radiation, the percentage of UV depends in a complex way on several parameters, such as: temperature, humidity, wet time, length of cloudless exposure period, wind direction and speed, latitude and angle of exposure. In order to estimate the amount of UV radiation based on global radiation data, the best practice is to measure radiation in the exposure radiation. Finally, the angle of incidence of the radiation has a relevant influence on the degradation rate of the exposed materials.

Furthermore, unique climatic and operating conditions can be encountered in specific geographical zone; an example is offered by Southern Africa. In this area, in fact, the presence of Antarctic ozone hole and powerful industrial pollutants leads to a particular severe environment. [Sookay et al., 2003] Tropical countries in equatorial area are characterized by high average annual temperature, elevated humidity values and relatively constant UV dosage. These climatic conditions are regarded as very detrimental for the performance of FRP strengthened structures [Liew and Tan, 2003]. Wide temperature variations possibly encountered under normal service conditions can adversely affect the FRP/concrete bond, since the coefficient of thermal expansion of epoxy resin is of about one order of magnitude greater than that of concrete. Data recorded by the National Climatic Data Center, Asheville, N.C., showed that over the past 50 years several regions in Florida experienced significant temperature changes, up to temperature range in excess of 55°C. [Sen et al., 1999] Finally, in many cold climates, such as those found in Canada and in the Northern U.S., seasonal and daily temperature variations have the potential to cause numerous freezing-thawing cycles than can, again, result in different thermal expansion in the FRP reinforce and the concrete substrate. Environmental data for Toronto, for example, indicates the potential for approximately 15 to 35 freeze-thaw cycles annually. [Bisby and Green, 2002]

Many are the examples of test conditions reproducing those encountered by composites for constructions during their service life. Freeze-thaw cycles under calcium chloride ($CaCl_2$) can be employed to simulate the deleterious effect of the deicing agents that are dispersed onto highways to safeguard against ice-formation, while wet-dry cycles under sodium-hydroxide (NaOH) can be used to simulate a condition of naturally-occurring alkalinity that arises from hydrated cement products, known to etch, spall, crack and dissolve the resin-matrix. [Davalos et al., 2004]

Adhesives and Primers Based on Cold-Cured Epoxy and Vinyl Ester Resins

The use of externally bonded FRP composites is a recent and attractive technique for the rehabilitation of deteriorated and under-strength concrete structures. FRP can be applied following two different procedures: a) the precured FRP prepregs are adhesively bonded as prefabricated strips to the concrete substrate; b) the composite is applied through a wet lay-up of fabrics directly onto the substrate.

The application of prefabricated strips ensures the use of precured factory produced materials, through industrially controlled processes, thus achieving a high level of uniformity for the final product and better performance. On the negative side, prefabricated FRP is less flexible and adaptable for unpredicted configuration that can be form in the field application. Moreover, the application of a precured FRP to a concrete substrate is carried out by means of a cold-curing thermosetting adhesive. This, therefore, implies the introduction of an adhesive interphase between the already cured FRP and the substrate.

The use of the wet lay-up technique provides enormous flexibility, since the pre-impregnated fabric can closely follow the geometrical configuration of the structure to be rehabilitated. Moreover, the bond between the FRP and the concrete substrate is achieved with an adhesive resin which is very similar to the matrix for the composite. Therefore, this adhesive interlayer is essentially part of the FRP itself by forming a resin continuum between the FRP and the concrete substrate. The lack of careful control of curing, however, leads to a significant higher level of variation in performance.

In both the described techniques the durability of a cold-cured resin, used as adhesive in the first case and as matrix/adhesive in the second, plays a fundamental role to assure the overall integrity of the rehabilitate of structure.

In general, the polymer adhesive and the matrix for the composite are regarded as the weak components in a FRP system, since it can undergo both chemical and physical degradation by the environmental conditions and mechanical stresses. Thermosetting resins are almost exclusively used as adhesives and matrices of FRP for structural applications. The most common resins used are epoxy or unsaturated resins, such as polyester or vinyl esters. These materials are chosen since they satisfy the requirements of an appropriate FRP matrix (low viscosity of the uncured resin, high reactivity on curing, no release, or very small, volatile products).

Among the polymeric adhesives and primers employed in the field of civil engineering, epoxy resins are without a doubt the most used for their excellent properties. Epoxy resins can be formulated into low viscosity systems which "cure" (i.e. form cross-links throughout the structure and, as a consequence, harden) at room temperature with a minimal shrinkage during curing. When correctly formulated and cured, epoxy resins exhibit a good combination of mechanical properties and chemical resistance towards environmental agents. Compared to alternative resins, cured epoxy systems are known to have excellent adhesion to a broad range of substrates and reinforcing materials. Different environmental factors have been found to have a deleterious effect on the durability of epoxy resins, such as moisture, freeze-thaw, temperature cyclical variations, alkaline environment, and ultraviolet radiations. The presence of humidity in the air is probably the most harmful environment that can commonly be encountered by epoxies used as adhesives for civil engineering applications. The sorption of water can greatly affect the physical properties of this thermosetting polymer and its composites. Water molecules bind with resins through hydrogen bonding. In this way, water is able to disrupt the interchain Van der Waals forces inside the network producing an increase of segmental mobility. [Zhou and Lucas, 1999, a)] As a consequence, the absorption of limited amounts of water can be regarded as beneficial in terms of both improved toughness and static fatigue resistance of the cured resin. On the other hand, an excessive penetration of water is generally considered harmful, leading to a reduction in modulus and strength with a consequent marked unsuitable decrease of load-bearing capacity through plasticization effects. [Mays and Hutchinson, 1992; Shaw, 1994] It is well known that the good properties of epoxy resins usually undergo a considerable deterioration after a long period of immersion in water. [Antoon and Koenig, 1981]

Vinyl ester resins can be sometime preferred to epoxies as matrices for FRP composites used in civil engineering applications, due to advantages in terms of processing conditions, reduced curing cycles and lower costs. Like the epoxies, moreover, they are often used in applications which require high acid resistance and/or high strength and stiffness; its resistance against corrosion is reported to be even higher than that of epoxies. [Riffle et al., 1998] In addition, they possess relatively high ultimate failure strains and damage tolerance.

The use of unsaturated polyester resins as matrices for composites in the field of construction and rehabilitation industry is rather limited, due to the high shrinkage occurred during curing (up to 10% by volume) that severely restricts the volume of material that can be employed. [Mays and Hutchinson, 1992] Reservations have also been expressed regarding the suitability of polyesters in such applications because the poor creep resistance under sustained load, a high susceptibility to moisture and their bonding efficiency in damp or wet conditions and in particular to alkaline substrates.

Although both vinyl esters and polyesters are styrene-based matrices, cured by the free-radical polymerization of the reactive diluent (styrene) and the unsaturated groups, the positioning of these unsaturated groups is different in the two different resins, leading to a different susceptibility to moisture/aqueous attack. In the case of polyesters, the unsaturated groups are within the molecular backbone of the system itself, thereby resulting in a greater number of sites for potential reactivity with an aqueous medium. Vinyl ester systems, on the other hand, have unsaturated groups only at chain end, thereby greatly reducing the possibility of deleterious reactions. [Zhang et al., 2000] Finally, both vinyl ester and polyester styreneated resins are particularly sensitive to UV radiations and should be stabilized and protected to infer a long-term resistance to this type of radiation. [Lesko et al., 2001]

Apart from specific requirements, all resins used for consolidation and restoration of concrete should be compatible with the original substratum, able to develop an adequate adhesive strength to this substratum, not harmful for operators, long lasting and, as far as possible, reversible. [Laurenzi Tabasso, 1992] In particular, the weathering and durability of products are very complex problems: the lifetime acceptable for normal utilization should be in the order of ten's years. It is well known that the final performances of the cured thermosetting resins (such as adhesive strength, heat and chemical resistance) depend mainly on the conditions in which they have been cured and in which they are used (temperature, presence of water, presence of external agents). The lifetime of an epoxy adhesive, for instance, will depend on the base components selected, on the conditions chosen for the cure (if possible to choose them) and on the conditions occurring during their service.

For economic and practical reasons, the thermosetting resins used in civil engineering applications must cure and harden at ambient temperature. Providing any kind of heat sources, over the large areas required for the described applications, is very difficult and prohibitively expansive.

In the case of cold-curing epoxies, aliphatic amines or polyamines are employed as curing agents, being they able to induce cross-linking reactions even at low temperatures. [Riviere, 1992; May, 1989; Mays and Hutchinson, 1992] One of the main drawbacks in the use of such curing agents, resides in the incomplete cure which results in poor performances of the cured systems, particularly with respect to the glass transition temperature. Incomplete conversion, furthermore, results in a lower resistance to hydrolysis and degradation of properties.

Referring to vinyl ester resins, an increase of the content of styrene, employed as reactive diluent in quantities ranging between 20 and 60%, can result in an increase in shrinkage, with the possible formation of micro-cracks and to the formation of micro-gels in the bulk, resulting in micro-inhomogeneities and incomplete polymerization. High levels of styrene lead, on the other hand, also to increases in hydrophobicity, thus in an decrease in the level of moisture absorption. [Rivera and Karbhari, 2002; Chu et al., 2004]

Effects of the Temperature

Due to the low glass transition temperature (Tg) of cold-cured systems, the cyclic variations in environmental temperature could cause drastic changes in their mechanical properties. The glass transition temperature is the temperature, or the range of temperatures, at which polymeric resins change from a rigid, glass-like state to an elastomeric-like state, with a consequently dramatic decrease in modulus and also affect the adhesive properties. The cross-linking of epoxy systems with aliphatic amine curing agents, suitable for low temperature curing, leads to systems possessing low Tg. In practice the Tg values are only few degree (max 10-20°C) higher than the curing temperature used, generally lying in the range 35-60°C [Mays and Hutchinson, 1992], which is not much higher than the maximum service temperature for many civil engineering applications. Even if the air temperature in Europe usually does not exceed 40°C, the temperature of the structure in direct contact with the surface irradiated by the sun can be higher than the Tg of the resin of the adhesive used. This situation becomes even more dangerous when cold-cured adhesives are used to bond metallic structures. In steel bridges in the UK, for example, temperature rises to 60°C or even 65°C may occur locally. [Mays and Hutchinson, 1992] Moreover, the Tg of the adhesive can decrease to values even lower than ambient temperature as a result of the absorption of water (atmospheric moisture) acting as plasticizing agent.

Frigione and coworkers (Frigione et al., 2000; Frigione et al., 2004, a)] studied how the properties of cold-cured epoxy resins, for which 30-40°C are the maximum temperatures recommended in service, can change when these temperatures are approached or exceeded for prolonged periods of time and if the modification of properties as a consequence of variations in temperature is or not reversible. A large deterioration in flexural properties was observed when the resins work at temperatures higher than 40-50°C, that is around the glass transition temperature (Tg) of the resins (ranging from 46° to 59°C), even if they are heavily filled (40% in weight) with an inorganic filler. The decreases in strength and elastic modulus at 50°C, for example, were measured to be by 38% and 53%, respectively, from the values at room temperature for an epoxy resin with a high Tg, i.e. 58°C. For another epoxy adhesive, having a Tg of 52°C, the reductions in both modulus and strength were even higher than 90%. It was demonstrated that the adhesives recover almost completely their properties when the service temperature is lowered below the glass transition temperature.

The reversibility of thermal treatments, performed on a commercial epoxy adhesive employed in the restoration of the Basilica of S. Francesco in Assisi, Italy, was investigated in a similar study. [Frigione et al., 2004, b)] The temperature chosen for the thermal treatment was 50°C, slightly lower than the estimated Tg, around 57°C, for the system cured for prolonged times (up to 4 months) at room temperature. The effect of this treatment was an increase of the Tg of the resin, which reached the value of 66°C after a 28 days thermal treatment. This "post-curing" process, moreover, led to an increase in flexural strength (by about 32% after a 28 days thermal treatment), in strain at break, which was more than doubled after 28 days. An increase in maximum strength as a consequences of post-curing process were already reported in literature. [Ellis, 1993; Meyer et al., 1995]

The mechanical (flexural and tensile) behavior of a commercial vinyl ester adhesive, cured at 23°C for 15 days and possessing a Tg of 59°C, resulted similar to that evidenced by epoxy adhesives when the service temperature was increased. [Aiello et al., 2003; Frigione et al., 2004, a)]

The exposure of epoxy resins to elevated temperatures brings about changes in properties, owing to additional cross-linking. An epoxy resin cured with methyl tetrahydrophtalic anhydride (MTHA) with a Tg of about 64°C, was evaluated. [Bellenger et al., 1990] Samples of the cured system were exposed at 140°, 160°, 180° or 200°C for 1500 hours. The spectroscopic, mechanical and thermal studies performed on the post-cured samples revealed the occurrence of thermal degradation through oxidative reactions. An increase of Tg was always registered, which was higher on the surface of the samples (up to 100°C) than in the core of the material, where the maximum Tg reached about 73°C. The effects of thermal treatments were reflected also on flexural properties. The flexural fracture, initially ductile, became brittle after a limit exposure time, which decreases by increasing the test temperature.

The behavior of a thermosetting resins for composites in relation to the temperature can be revealed also from studies performed on adhesives. When a thermosetting adhesive is employed to join metallic, glass and ceramic adherends, the adhesive can be exposed to a temperature higher than its Tg; this can weaken the joint and lead to cohesive failure, i.e. failure within the adhesive. [Comyn, 1990]

The effect of service temperatures close to the Tg's of the epoxy adhesives on their adhesion strength to several materials was examined in different studies. [Brewis et al., 1982; Brewis et al., 1983; Van Gemert and Vanden Bosch, 1986; Franke, 1986; Tadeu and Branco, 2000; Aiello et al., 2002] In all cases, a substantial decrease in the adhesive strength at around the Tg of the epoxy resin was found, with a more frequent occurrence of slip at the interface adhesive/adherent rather than a cohesive fracture within the adhesive. The effect of the temperature was confirmed to be in some extent reversible, since the results indicated that a limited number of severe thermal loadings (up to 75°C) do not have a significant effect of the effectiveness of the bond, when the temperature was lowered to ordinary ambient values. [Van Gemert and Vanden Bosch, 1986; Aiello et al., 2002]

As an example, the durability of reinforced concrete (RC) beams, externally strengthened on the tension side with single layered CFRP plates (supplied as heat-cured prepregs), was analyzed by exposing the system to different harsh environments. [Grace et al., 2002; Grace and Grace, 2005] The adhesive used to bond CFRP plates to concrete elements was a structural cold-cured epoxy, with a glass transition temperature by about 60°C. To examine the effect of dry-heat conditioning on CFRP strengthened beams, the specimens were aged in a chamber at 60°C up to 10000 hours. Four-point loading tests were performed on heat treated and untreated (control) strengthened beams. The results showed that, even if the temperature selected for the thermal treatment was equal to the Tg of the adhesive resin, the heat treatment did not affect significantly the bond strength for mechanical tests performed at ambient temperature. The reduction in the strength of the beam due to dry-heat aging was only around 13%.

Effect of Moisture and Liquid Water

Epoxy matrices can absorb substantial amounts of water when exposed to liquid water or moisture from the environment, due to the presence of polar groups able to attract water molecules. The effects of the exposure to moisture is the deterioration of physical and mechanical properties.

The performance in aqueous environments of commercial epoxy adhesives and primers was studied by subjecting samples, previously cured for 20 days at ambient (23°C) temperature, to immersion in distilled water for different times. [Frigione et al., 2004, a); Frigione et al., 2006] Thermal and flexural tests were used to determine the Tg and the mechanical properties of exposed and control specimens of epoxy adhesives. A reduction in the glass transition temperature and in the stiffness at short immersion time was found for all the adhesives analyzed, attributed to plasticization effects. A subsequent slightly increase of these properties was registered for prolonged immersion, probably due to a continuation of curing reactions. The maximum reduction in Tg was measured for the adhesive with the highest initial Tg. In saturation conditions, moreover, the adhesives with initial lower Tg values almost recover the initial Tg, while that with initial highest Tg showed a decrease in this property by about 12°C. The observed changes in strength followed a similar trend, i.e. rather insignificant at short immersion time and reductions around 10% at longer exposure times for the resins possessing lower initial strength values. Higher decreases in strength were observed for adhesives with initial higher Tg value, i.e. after one month of immersion in water the reduction in maximum stress was by about 17% and the adhesive retained this value up to saturation condition. The strain at break was decreased by about 20-50% as a consequence of immersion in water.

The behavior of other commercial thermosetting matrices for composites (i.e. epoxy cold-cured primer and adhesive), employed for the restoration of concrete structures, was analyzed in relation to the most common service conditions. [Frigione et al., 2004, a); Frigione and Sciolti, 2006] The results of tests performed on the epoxy primer after immersion in water showed an initial decrease in flexural strength and modulus (about 50% and 70%, respectively) after 50-60 days of immersion, corresponding to the saturation time. Then the strength increased again, to values similar to those relative to the control samples. Similarly, for longer immersion periods the modulus increased again, with a reduction by about 15-20%. Finally, the strain at break increased up to 50% depending on the saturation time. At longer immersion times, on the other hand, the strength was similar to that of the unaged samples. A peculiar behavior was observed for the epoxy adhesive. It did not reach the saturation conditions, until 25 weeks of immersion. The strength increased by about 40% after only 2 weeks of immersion and retained this value even at longer immersion time. After 22 weeks in water, the increase in strength was by about 10%. The flexural modulus and the ultimate strain, after an initial increase by about 25% at short immersion times, remained roughly constant. The latter results would exclude the degradation of the resin, produced by the long immersion in water, and the consequent formation of microcracks, that could have explained, on the other hand, the not attained saturation condition.

The effects of water, in liquid or vapor state, on a commercial epoxy adhesive, employed in the restoration of the Basilica of S. Francesco in Assisi (Italy), were evaluated in some details. [Frigione et al., 2004, b)] The hygrometric treatments were conducted by immersing the samples in water at 23°C or keeping them at a controlled humidity level equals to 100%

R.H. The results showed that the exposure to water, whether in liquid or vapor state, produces similar effects, since the amounts of water absorbed in both experiments was very similar. A decrease in Tg of the samples subjected to both treatments for short times was also confirmed in this study. At longer exposure times, i.e. when saturation is reached, the Tg increased again to value similar to one for the unaged sample. Again the continuation of cure was considered to be the cause. The results of the flexural tests showed that both hygrometric treatments produced small decrease in flexural modulus (not exceeding 15% in both cases) and negligible variations in maximum strength and ultimate strain, for exposure times of approximately one month.

The results obtained in the mentioned studies were in good agreement with the results reported by other authors. Referring to the effect of immersion in water on the thermal properties of epoxy adhesives, it was found that relatively short time of exposure produces reversible plasticization effect, i.e. a lowering of the Tg. [Li et al., 1990] This effect is particularly worrying for cold-cured epoxies whose typical glass transitions, when dry, is not much higher than the possible service temperature. Hence, the need arises to select adhesives whose Tg's do not drop substantially with water sorption or to ensure almost dry ambient conditions.

The increase in Tg after a longer immersion time is undoubtedly due to additional cross-linking during exposure to water. Since water can cause plasticization of the resin [De'Neve and Shanahan, 1993], so that the lowering in Tg allows the network chains to become more mobile, which permits a certain degree of post-curing. [Marshall et al., 1982] This was confirmed by experimental results, showing higher values of Tg after long immersion time and higher exposure temperatures. [Zhou and Lucas, 1999, b)] In particular, the Tg was found to decrease in a first stage of water absorption process until it reached a minimum in a second stage. Afterwards, the Tg slightly increases with exposure time. The greatest reduction in Tg as consequence of the absorption of water seems to be related to the initial Tg values. The higher the initial Tg value the greater the reduction in Tg. Moreover, the initial Tg value influences accordingly also the increase in Tg after the first immersion period.

The effect of the immersion in water for a prolonged period on the mechanical properties of mild-cured bisphenolic epoxy adhesives (Tg = 76°C) was manifested in an initial increase in ultimate tensile strength (up to 21% after a 3 months immersion) followed at longer immersion times (i.e. 5 months) by a decrease to values similar to that of the unaged polymer. [Kajorncheappunngam et al., 2002] Similar results have been found by other authors also for cold-cured bisphenolic epoxy matrices for composites (curing temperature = 20°C). [Rinaldi and Maura, 1993] The initial increase in the ultimate tensile strength was again explained in terms of an increase in cross-link density. [Gupta et al., 1985] The subsequent reduction in strength was attributed to the long immersion in water. From the same study, the Young's modulus of the aged epoxy was found to be marginally lower than that of the control samples. The reduction in modulus continued with increasing the immersion time (after a 5 months immersion the reduction approached the 18%). The elongation at break, moreover, tended to increase initially (up to 76% after a 3 months immersion), but, later, the material became brittle, showing final increase of about 30% with respect to unaged samples after a 5 months immersion. It has been reported that a reduction in the failure strain can be regarded as a clear and sensitive indicator of polymer degradation. [White and Turnbull, 1994]

The effect of the presence of water on adhesive resins is often reflected on the behavior of the joint. In the case of steel-polypropylene joint, employing a cold-cured epoxy adhesive, lap-shear specimens were kept in a climate of 30°C and 60% R.H. and tested after 0, 2, 6, 8, 12, 24 and 36 months of exposure. [Kollek, 1990] The results showed that in the first two months of exposure the lap-shear strength increased, attributed to the post-curing of the adhesive and to the ingress of water at the edges of the glue line, which lowers the stress concentration at these edges. Thereafter, the lap-shear strength was found to decrease, reaching a minimum after about 8 months. This was attributed to the water distribution becoming more even along the glue line and causing the weakening of the adhesive as a result of plasticization. Then, it increased again, after about 12 months, at which point the adhesive joint reached an equilibrium with the environment over the entire joining area. After 12 months a slow decrease of strength was observed, due to the long-term aging effects. Finally, after 24 months, a rapid corrosion occurred, leading to total failure of the joint. Similar results were also reported for the same joint when exposed up to 36 months to similar environmental conditions.

The effect of liquid water at 35°C on a cold-cured (23°C) epoxy adhesive employed to join an aluminium alloy to a carbon fiber reinforced phenolic composite, was studied. [Popineau and Shanahan, 2006] Some tests were performed in torsional mode on tube made of aluminium glued using the epoxy to composite plates. Aluminuim plates glued to composite plates were tested using wedge test configuration. A continuous decrease, up to 80-90%, in energy of adhesion was found concurrently with a similar change in ultimate shear bond stress with increasing immersion time. The dramatic loss in adhesion performance was reflected by a similar drop in mechanical properties of the polymeric adhesive, up to 20 days of immersion. X-ray surface analysis proved that the decrease in adhesive strength was due to a progressive aluminuim/epoxy adhesive interfacial debonding with immersion time, caused by water ingress in the adhesive at and/or near the interface.

The effect of short-term and medium-term exposure to liquid water on the strength of the bond developed between epoxy resin adhesive and concrete was investigated in two successive studies. [Aiello et al., 2002; Frigione et al., 2006] Three commercial epoxy adhesives (Tg's ranging from 46°C to 58°C) were applied to concrete samples and cold-cured for 20 days before immersion in distilled water. In the first paper, was reported that the presence of water induces a certain reduction in the bond strength (15-25% depending on the strength of the concrete used) even after a 48 hours exposure to water. In the second work, the adhesion strength exerted by different cold-cured epoxy adhesives to concrete was found to be reduced, generally by about 30% after one month of immersion in water. A weakness of the resin/concrete interface was revealed in all cases, irrespective of the duration of the immersion in water.

The effect of prolonged immersions in water (up to four years) on adhesive/concrete joints was analyzed in a previous research work. [Franke, 1986] Two epoxy resins were chosen as adhesives to bond concrete cylindrical elements; the formed specimens were exposed to a wet ambient and, after different time spans, subjected to shear compression tests. The adhesion strength of both resins was found to decrease by about 20% after two weeks immersion. At longer exposure times, one adhesive showed a continuous decline in bond strength up to 30%, another adhesive showed a slight increase in strength, to finish with a bond strength loss of only 10% from the initial value. Also in this case, the adhesive/concrete interface was found to be the region in which the failure preferentially occurred.

A commercial cold-cured epoxy adhesive, possessing a Tg by about 60°C, was employed to bond a heat-cured single layer carbon fiber reinforced polymer (CFRP) plate to reinforced concrete (RC) beams. [Grace et al., 2002; Grace and Grace, 2005] The durability of the RC strengthened beams was analyzed under several environmental conditions by four-point loading tests. It was found that the long term exposure to humidity is the most detrimental factor affecting the bond strength of the epoxy adhesive between the CFRP plates and the RC beams. CFRP strengthened beams exposed for 10^{+4} hours (i.e. almost 14 months) to a moist environment (100% R.H., 38 ± 2°C) showed a considerable reduction in strength relative to the un-exposed specimens. The strength was found to continuously decrease with increasing the exposure time, up to a 32% reduction after 10000 hours. Debonding and onset of delamination were the major modes of failure observed.

Considerable information on the durability of epoxy adhesives can be derived from studies performed on coatings for concrete structures. First, the ambient conditions at which the cure takes place have been found to be the main factor for the effectiveness of the adhesion between the resin and the concrete substrate. [Littmann, 1999] While the number of adhesion failures increases for the case of wet specimens, the average adhesion strength is less influenced by the presence of water on the samples. When the cure is performed at 23°C significant better adhesion characteristics were found with respect to those cured at 8°C. The storing conditions were found to be equally important, both wet and humid environments caused a decrease in adhesion strength. Finally, the duration of storage conditions seemed to be of lower importance, irrespective of the humidity, and the adhesion properties even appeared to improve with time. A wide spread range of results was, however, observed in most cases.

Several epoxy formulations, used for coatings on concrete, were cured under laboratory climate conditions (23°C) and subsequently immersed in tap water for prolonged time. [Wolff et al., 2006] The absorption of water, measured as gain in weight, showed different features in relation to the chemical formulation of each epoxy system analyzed. In particular, a higher increase in water absorption (higher than 5% wt.), continuously increasing even after a 6 months immersion time, was noted for all commercial epoxy formulations. This behavior was attributed to the high amount of benzyl alcohol (about 20% wt.) present as diluent inside the system. The majority of the formulations tested were characterized by an amount of water uptake around 2-3% wt. They reached the saturation condition after storage times of about 3-4 months. Formulations based on different amine hardeners and containing amounts of solvent (e.g. ethyl alcohol) ranging from 3 to 35% wt. also showed this tendency. Other formulations showed a small decrease of weight after an initial increase, probably caused by leaching of soluble additives not bounded to the resin network. The epoxy formulation which showed more prominently this behavior had a content of benzyl alcohol in the hardener of about 35% wt.

There is a very poor understanding of the mechanism of aging and environmental degradation for vinyl ester resins cured at moderate temperatures, especially in presence of liquid and vapor water.

Vinyl esters consist of low molecular weight polyhydroxyether chains with unsaturation (reactive groups) at chain ends. The addition of styrene, in the monomeric form, affects the proceeding of the curing reaction, resulting in an increase in the final performance (mechanical, Tg, etc.). As already explained, an increase in styrene content results in an intrinsic increase in hydrophobicity, volumetric shrinkage and the likelihood of incomplete polymerization, which can also produce a degradation of the properties of the resin and of its adhesion characteristics. Furthermore, the hydrolysis of ester groups can result in the formation of carboxyl acid groups, resulting in further deterioration due to autocatalysis. [Apicella et al., 1982] Incomplete polymerization leads to changes in properties with time, induces lower heat stability (low Tg), lowers resistance to hydrolysis, and induces a greater susceptibility to swelling when exposed to solvents. [Apicella et al., 1983]

A study on diffusion of different liquids in mild-cured composite thermosetting matrices (Tg's = 80-100°C) showed that the equilibrium mass uptake of liquid water in vinyl ester resins was lower than in epoxy resins. [Chin et al., 1999] Vinyl ester resins are less hydrophilic relatively to epoxy resins containing a lower concentration of polar functional groups than epoxies. On the other hand, the diffusion coefficient calculated for vinyl ester resins was higher than that found for epoxy resin. This was confirmed by the higher permeability to water of vinyl ester resins relatively to that of epoxies, since permeability (P) is a function of solubility (S) and diffusion (D) coefficients, i.e. $P = S \times D$.

Effect of Freeze and Freeze-Thaw Cycles

The effect of freeze and freeze-thaw conditions on the properties of cold-cured neat resins has hardly been analyzed. Some indications, however, can be derived from studies performed on epoxy adhesives used to join different materials. In such cases, the deterioration of the bond strength, resulting after repeated freeze-thaw cycles, is often attributed to the different coefficients of thermal expansion of the adhesive and of the constituents of the joint. [Bisby and Green, 2002]

The effect of dry freeze-thaw cycles on the adhesion strength of a two-part epoxy adhesive employed to joint concrete elements to glass fiber reinforced polymers (GFRP) plates, obtained by the pultrusion process, was analyzed. [Mukhopadhyaya et al., 1998] The cycles were realized in air, with the temperature varying from -17.8°C to +20°C, performing two cycles a day, up to 450 cycles. Double lap shear tests were employed, in order to ensure that the damage brought about by the test regime occurred at the interface. In this way, the effect of the harsh environment was mainly concentrated in the adhesive layer. The unfavorable exposure conditions were found to produce severe damage to the GFRP plate-concrete-adhesive interfaces. In particular, higher slips at the interface occurred in the exposed specimens when compared to the untreated control samples, suggesting that a severe degradation took place in the adhesive.

The effect of freeze-thaw cycling on reinforced concrete beams externally strengthened with a pultruded CFRP sheet, was also evaluated. [Green et al., 1998; Bisby and Green, 2002] The carbon fiber composite plate was applied to concrete beams through a cold-cured epoxy adhesive. The resulting strengthened beams were subjected to repeated freeze-thaw cycles (from 0 to 300 cycles), consisting in freezing in air at -18°C for 16 hours and thawing in water at 15°C for 8 hours. They were, then, tested to failure in four-point bending. The results of flexural tests indicated that ultimate load and midspan deflection as well as average bond stress at failure increased by increasing the number of freeze-thaw cycles. The failure mode for CFRP plated beams consisted of a peeling away of the layer of concrete between the FRP plate and the internal reinforcing steel, often referred to as shear peeling. After a small number of freeze-thaw cycles, the failure surface did not significantly change as compared with the room temperature specimens, showing failure within the substrate concrete and remaining undamaged the adhesive-concrete bond. After 150 cycles, on the other hand, peeling away of the concrete between the FRP plate and the internal reinforcement was accompanied by the formation zones of debonding along the epoxy-concrete interface and failure within the FRP plate itself. After 300 cycles, finally, larger portions of the failure surface featured epoxy-concrete debonding and failure within the FRP laminate. The authors stated that the observed changes in failure mode, as a consequence of freeze-thaw regime, did not correlate with the increased values of the failure load of the beams calculated with increasing numbers of freeze-thaw cycles. They hypothesized a redistribution of load at high load levels as consequence of freeze-thaw exposure. This would damage the epoxy adhesive and reduce the effect of stress concentrations at the FRP-concrete interface, possibly changing the location of the failure surface as a result.

A cold-cured epoxy adhesive (Tg = 60°C) was used to bond CFRP plates to reinforced concrete (RC) beams. [Grace et al., 2002; Grace and Grace, 2005] Single layered CFRP plates were supplied as heat-cured systems. To determine the performance of the strengthened concrete elements under freeze-thaw regimes, the beams were exposed to 350 and 700 freeze-thaw cycles. The temperature in each cycle varied between -17.8° and 4°C; air was used to freeze the beams while water was employed to thaw them; each freeze-thaw cycle took four hours. The strengthened beams were then tested by a four-point loading method. For comparative purposes, some control specimens were tested at the same time. The exposure to freeze thaw cycles caused a small reduction in load carrying capacity of beams, i.e. about 3% and 9.5% after 350 and 700 cycles, respectively. The result suggested that neither of the two cold-cured epoxy resin was affected by the freeze- thaw regime, nor was the CFRP plate.

Effect of Aqueous Solutions and Chemicals

In a recent paper, the sorption behavior of a cold-cured (cured at room temperature for 24-48 hours) epoxy adhesive under complete immersion in distilled water and sodium chloride solutions, was investigated. [Kahraman and Al-Harthi, 2005] It was found that the epoxy adhesive absorbs a larger amount of water upon exposure to distilled water than when exposed to aqueous NaCl solutions. The higher the concentration of the salt in the solution, the less was the amount of water absorbed by the adhesive. This effect was explained by the reverse osmosis mechanism. The rate of diffusion in the adhesive immersed in a test solution was found, however, proportional to the salt concentration in the solution. It was, therefore, hypothesized that the concentrated salt solutions somehow enhanced the formation of microcavities in the material which would act as reservoirs for the infused water.

The durability of externally strengthened RC beams was analyzed in aggressive environments, such as salt-water and alkali solution. [Grace et al., 2002; Grace and Grace, 2005] The strengthening was performed by applying single layered CFRP plates (supplied as heat-cured systems) on the tension side of the beams. To this end, a structural type cold-cured epoxy was employed, having a glass transition temperature around 60°C. The specimens were immersed in alkaline and salt solutions at 23°C. Four-point loading tests were performed on beams after different immersion times, as well as on control specimens. It was found that the exposure to both saline and alkali solutions did not affect the bond strength of the epoxy adhesive. The load carrying capacity of beams even increased, especially for short-term exposure (i.e. 3000 hours).

The durability of steel elements strengthened with CFRP plates when exposed to different environments, was studied. [Colombi et al., 2005] Steel plates were reinforced using pultruded CFRP plates (purchased as heat-cured systems); the adhesion between CFRP and steel was established by applying a thixotropic epoxy resin, for which a curing cycle of 7 days at 35°C was recommended by suppliers. The final Tg of the resin was 62°C. "Double-doubler joints" were prepared using two CFRP strips bonded to two steel plates by the epoxy resin. The specimens were designed to reproduce the geometry of a real application. In order to evaluate the effect of high moisture content in presence of salt on the bond strength between CFRP and steel, a joint was aged for 3 weeks in a salt spray fog, with a relative humidity of 95% and a salt concentration of 5% (NaCl). The aged specimen was tested under tensile loading. For comparison purposes, an untreated specimen was tested in the same conditions. The efficiency of the adhesive bond was found to decrease as consequence of the severe exposure. The reduction in ductility, measured for the aged specimen with respect to the control, was by 22%.

The durability of different formulations based on epoxy resins and used for coatings on concrete was analyzed in relation to the conditions within the concrete. [Wolff et al., 2006] To this end, the different formulations, previously cured at 23°C, were stored in an artificial solution reproducing the conditions found in concrete, with a pH of 13.8 for 6 months. The amount of absorbed solution of each formulation was found to be very similar to the amount of pure water absorbed by the same formulations, earlier reported. Moreover, no loss of resistance was recorded in most of the analyzed epoxy formulations as a consequence of the immersion in the alkaline solution.

Effect of Combined Agents

Thermo-hygrometric treatments were performed on a commercial epoxy adhesive, employed in the restoration of the Basilica of S. Francesco in Assisi (Italy), in order to simulate a typical indoor exposure inside the cathedral. [Frigione et al., 2004, b)] The treatments were performed in a climatic chamber, varying the temperature between 23° and 45°C and the humidity level between 50% and 90% R.H.. It was previously found that the effect of the temperature was to increase the Tg, due to the continuing of cross-linking reactions, whereas the effect of the high level of humidity was a reduction in Tg, through plasticization. The combined effect of temperature and moisture caused a slight increase in Tg, immediately after the end of the combined thermo-hygrometric treatment. This was, however, lower what was expected from an additive effect of the two treatments, i.e. thermal and hygrometric. The Tg, however, increased when the measurements where made on samples after the hygrometric treatment stored at ordinary ambient conditions for 1 month, or longer. The latter increase in Tg was explained in terms of a slow evaporation of the water absorbed during the treatment. The mechanical properties, on the other hand, were mainly affected by the duration of the treatment at a high temperature. The results obtained immediately after the end of these combined treatments were, in fact, similar to those found on samples aged in oven at 50°C, a part from the large increase in strain at break, which was attributed to plasticization effects.

The effect of the two common environmental agents (heat and moisture) on the properties of three commercial epoxy resins, used as adhesive, putty and primer, for the manufacture of an hand lay-up composite, was studied by subjecting the samples to a treatment of about six months in a climatic chamber at 40°C and 90% R.H., which corresponds to a warm climate achievable in Mediterranean regions. [Aiello and Sciolti, 2005; Frigione and Sciolti, 2006] The results of tensile tests performed before and after the treatment confirmed the considerable effect on the mechanical properties. The mechanical tests on the epoxy primer could not be carried out because the conditioned specimens were completely damaged after the treatment performed. Significant variations in mechanical properties were registered for the epoxy putty, consisting of an increase in modulus of elasticity in the region of 90%. For the epoxy adhesive, the tensile strength and the modulus of elasticity decreased by about 70% and 17%, respectively.

The same epoxy primer, previously studied in isolation, was applied to bond a natural stone, i.e. Leccese ashlar, to CFRP and GFRP sheets (containing unidirectional fibers) supplied as systems cured at high temperatures. [Aiello and Sciolti, 2005] Specimens of stones reinforced with these two FPR's were subjected to the same thermo-hygrometric treatment performed on the single materials. They were then tested in a double-lap shear mode, and the results compared with those obtained on unexposed samples. While for unconditioned specimens the failure occurred always with a complete detachment of the composite sheets together with a thin layer of calcarenite stone, the failure behavior of the conditioned samples reinforced with GFRP was affected by the degradation of adhesive. It was often observed the sheet broke with formation of cracks parallel to the fibers, or by delamination of the sheet with the detachment of a very thin layer of stone. On the other hand, the bond strength measured on specimens reinforced with GFRP seemed to be unaffected by the previous conditioning. The treatment appeared to be more influential on the bond strength of the actual specimens reinforced with CFRP. In these samples, high failure strain values at a fixed load applied to the sheets were also registered after the conditioning. This was attributed to a decrease in modulus of the adhesive as a result of the thermo-hygrometric treatment.

The influence of aggressive exposure conditions on the behavior of an epoxy adhesive used to bond concrete/glass fiber reinforced polymers (GFRP) joints, was analyzed by exposing the specimens to alternate wet-dry cycles in a saline solution and to a combination of wet-dry cycles in the same solution and dry freeze-thaw cycles. [Mukhopadhyaya et al., 1998] To this end, GFRP plates, previously heat-cured, were bonded to concrete elements employing a two-part epoxy resin as adhesive and cured at room temperature. Double-lap shear specimens were tested in order to examine the behavior at the interface. Some specimens were, then, immersed in 5% sodium chloride solution for one week, followed by drying in air for another week in ambient temperature (i.e. 18°C). After a total of 18 such wet-dry cycles, the samples were tested in double lap shear mode. Contemporarily, 18 cycles were performed on other specimens, one cycle consisting in one week immersion was followed by 14 freeze-thaw cycles in air, lasting another week. Dimensional changes occurred in the specimens subjected to the wet-dry cyclic exposure regime, even though no differential movements between the GFRP plate and the concrete was recorded. The failure mode was predominantly through "concrete shearing", indicating that even after this treatment, the adhesive used still provided a stronger adhesion bond than the shear strength of the concrete. However, a few zones were noted, where there was some loss of adhesion between concrete and adhesive. These zones were not observed in the case of the untreated specimens, indicating that the exposure affected to a certain extent the adhesion strength of the resin. In the case of specimens subjected to a combination of wet-dry cycles and dry freeze-thaw, very large differential movements between the GFRP plate and the concrete substrate were observed, when compared to the effects produced by the two treatments in isolation. The results clearly suggest that these two exposure regimes in the long run are likely to damage the integrity of the adhesive bonded specimens. The duration of exposure, on the other hand, was not long enough to affect significantly the strength of the adhesive joint. Referring to the failure mode, wide areas of adhesion failure at the interface between the concrete and the epoxy adhesive were observed, being these zones uniformly distributed all over the surface. The results, therefore, have demonstrated that this combined aggressive exposure is potentially very harmful for the integrity of the interface in the bonded joint, being able to inflict severe deterioration of the bond strength of the epoxy adhesive.

Double-strap joints, consisting in steel plates bonded to CFRP laminates, were analyzed after exposure to a harsh environment. [Colombi et al., 2005] The joints were made from pultruded CFRP strips (i.e. purchased heat-cured) bonded to steel plates with a thixotropic epoxy resin (recommended curing cycle: 7 days at 35°C). A combined treatment, consisting of 20 thermal cycles of six hours at -20°C and six at 50°C, plus 2 weeks in a spray salt-fog cabinet (relative humidity = 95%, salt concentration, NaCl = 5%), was performed on a steel/CFRP joint. The temperature chosen for the treatment was below the glass transition temperature of the epoxy adhesive, stated by the manufacturer to be equal to 62°C. Uniaxial tension tests were performed on both control and aged samples. The treatment performed produced a detrimental effect on the final performance of the adhesive in transferring loads between the steel plate and CFRP. A significant ductility reduction of the adhesive layer (about 50%) was measured. Moreover, the reduction was significantly higher also than that measured after the exposure to salt-fog conditions without thermal treatment, i.e. 22%.

The effect of different harsh aqueous environments have also been investigated on a cold-cured vinyl ester resin. [Sagi-Mana et al., 1998] The resin was cured at room temperature for two days. After this cold-curing stage, the degree of reaction (i.e. the percentage of conversion of reactive groups) was estimated to be 83% and the Tg of the resin around 60°C. Post-cure at 100°C for 12 hours was performed on some samples previously cold-cured, obtaining a final degree of reaction equals to 89%, i.e. still not fully cured. The calculated Tg for the post-cured resin was 117°C and, therefore, substantially higher than that measured on the cold-cured specimens. The increase in Tg after the post-cure can be attributed not only to the increase in the degree of reaction (from 83 to 89%) but also to the removal of low molecular weight plasticizing molecules (styrene). Both cold-cured and post-cured samples of vinyl ester resin were, then, subjected to different accelerated environmental aging conditions, i.e.: immersion in artificial seawater at 60°C up to 6 months; immersion in boiling seawater for 48 hours; and immersion in boiling distilled water for 48 hours. The samples were also exposed for up to 1000 hours to weathering tests, consisting of 8 hour cycles at 60°C with ultraviolet irradiation followed by 4 hours at 50°C with water condensing on the surface of the sample. All samples were then dried for two weeks at ambient temperature prior to testing. Both thermal analysis and mechanical tests, in a 3 point bending fixture, were performed on the samples subjected for several time spans to the different environmental conditions. It was found that the mere immersion of cold-cured samples in seawater at 60°C resulted in further curing and led to some extraction of low molecular weight substances, which resulted in an increase in Tg, to 104°C and 106°C after 15 and 30 days of immersion, respectively. Accordingly, the flexural modulus of the cold-cured resin increased, upon immersion in warm seawater, to reach values equal to those for unaged post-cured resin. After longer immersion time, on the other hand, the Tg slowly decreased, down to 98°C after 6 months of immersion. This was attributed to low levels of hydrolysis and to the presence of small amounts of residual water in the resin, even though the samples were dried for two weeks before the measurements. The post-cured samples were hardly affected by the immersion in seawater at 60°C, in terms of Tg and flexural modulus. No appreciable difference was observed, in any of the studied systems, between the effect of boiling seawater and boiling distilled water. Both resulted in a post-cure, as indicated by the notable increase (about 60%) in flexural strength of the vinyl ester samples. Cycles of ultraviolet radiations at 60°C and water condensate at 50°C, for up to 1000 hours, also caused post-curing effects, but the changes were lower.

Effect of Natural Exposition

The effect of natural weather was studied by exposing some epoxy adhesives to the natural weather conditions for up to 36 months. [Frigione et al., 2001, a); Frigione et al., 2001, b)] Samples of these adhesives were periodically tested, measuring their glass transition temperature and mechanical properties. Some exposed samples, moreover, were subjected also to a thermal treatment (24 hours at 50°C) before performing any thermal analysis and mechanical test. It was found that the natural aging conditions produced a cycling change of Tg for both adhesives, depending on the meteorological seasons. The changes in Tg, due to plasticization effects, are mainly affected by the presence of humidity and the fluctuations in temperatures. This apparently reversible behavior, in so far as the average Tg values are only slightly lower than the initial values, cannot be erased by the thermal treatment, which is capable of removing only a limited amount of sorbed atmospheric water. The exposure to natural weathering, moreover, causes also a very mild fluctuation in mechanical properties. The results obtained in these studies for a period of 3 year natural exposure could be reasonably extrapolated to longer periods in absence of extremely severe external agents (i.e. corrosive ambience, one of more than one hazardous agents).

In a similar study, a commercial two-component epoxy adhesive, cured at 23°C for 4.5 months (to reach the complete cure), was exposed for up to one year to natural aging. [Lettieri et al., 2006] Periodic measurements on the exposed samples confirmed the above stated fluctuating behaviour of the glass transition temperature in concordance with the cycling climatic conditions. A slight decreases in Tg values occurred during the wettest months, while higher Tg's were recorded in warmest season. The latter effect was again attributed to both the occurrence of post-cure and the evaporation of sorbed water. The maximum variations in Tg registered were between -7°C and +3°C. Similarly, the results of the flexural mechanical tests carried out on weathered samples revealed an initial decrease in flexural modulus and strength, concomitantly with an increase in elongation at break, which are all indicative for effect of plasticization process due to water absorbed from atmosphere.

The effects of exposure to external environment on the properties of epoxy adhesives used for joints with different materials (steel and concrete) were examined in several independent studies. [Ladner, 1983; Van Gemert and Vanden Bosch, 1986; Van Gemert et al., 2001; Frigione et al., 2001, c)]

In a pioneering study, corrosion and creep measurements were undertaken on concrete beams strengthened with steel plate bonded with an epoxy adhesive and subjected to outdoor aging for more than 10 years. At the end of the exposure period, no difference in static behavior (up to failure) was observed. [Ladner, 1983]

In a different study, no remarkable loss of strength of the epoxy adhesive used to bond steel cylinders to concrete plates was found after exposure periods of up to two years, while only a relatively small decrease (10-20%) was measured for longer exposition periods. [Van Gemert and Vanden Bosch, 1986] In any case, it has been found that the preparation of the steel surfaces plays an important role. When the surface of the steel component are only sawed and degassed, but not grit-blasted, the resulting cutting lines act as preferential paths for the water to penetrate between the steel and the adhesive, causing the formation of rust with a consequent severe decrease of the bond strength between adhesive and steel.

In another study, the durability of "glued connections" on aging in atmospheric conditions was assessed by tear-off tests performed on small steel cylinders bonded to a concrete surface (the deck of three bridges) using an epoxy adhesive. [Van Gemert et al., 2001] Some cylinders were glued on the surfaces that were permanently in shaded areas, while others were bonded on surfaces heated by sun radiation. Six cylinders were tested both in the summer and in winter period over a ten year period. Temperatures in wintertime ranged from –10° to +9°C, and in summertime from 14° to 26°C. No decrease in strength was measured on all the cylinders tested after 10 years but the epoxy bond between the steel plates and concrete failed after 15 years and 9 months, following a very hot day with maximum temperature of 37°C, as reported in the same study. The failure of the adhesive, whose Tg was about 60°C, was attributed to the high external temperature, which was considered to have increased further by the direct sun radiation.

A limited decrease (up to 30%) of the adhesion strength of epoxy adhesive-concrete joints was registered after natural aging over different periods of time, up to one year. [Frigione et al., 2001, c)] The degradation of the bond properties was attributed to the decay of the mechanical properties of the adhesive and the weakening of the interface, due to the absorption of water, and/or to an increase in the temperature. The amount of degradation was, however, lower than that resulting from exposure to a heat and humid ambient in laboratory.

Accelerated Tests

Field monitoring is recognized as the only effective method of assessing the short and long term behavior of materials. The combined effects of environmental factors and the variability of conditions often result in different rates, and mechanisms, of degradation. This makes it difficult, therefore, to obtain good correlations with controlled laboratory exposures. However, durability reliable studies require long periods of natural exposure and, therefore, accelerated laboratory tests will always be used.

Several standardized procedures dealing with the accelerated weathering conditions, are available. In these one or more weather like conditions are intensified to levels greater than those occurring naturally. For example, one of the frequently applied methods to accelerate the effect of exposition to water is immersion in boiling water. Another possible method is to employ significantly increased levels of irradiation and temperature for testing under simulated conditions in weathering cabinets. None of these standards, however, imposes rigid conditions on how an accelerated test should be performed. This is left to the experimenter to decide, even though there are some guidelines for choosing important test parameters. However, some accelerated environment conditions are often employed to produce effects which are not realistic and unrelatable to natural conditions.

When temperature is used to accelerate aging, most researchers make use of the Arrhenius rate equation for chemical processes. [Nelson, 1990] This involves conditioning and testing the materials at a number of different temperatures and the rate of degradation of chosen properties is determined at extrapolated service temperature (i.e. 23°C). There are several reservations, however, on the applicability of this procedure to cover a wide range of temperatures, due to the changes in the physical state of the material. In particular, the conditioning procedure at moderate temperature, say lower than the Tg of the resin, can have an effect on the structure of cold-cured resins. As a result, the conditioning treatment can change the mode in which the material degrades under actual service conditions. [Bank et al., 2003]

Owing to the difficulty to relate accelerated tests procedures to the performances of the materials in actual service [White and Turnbull, 1994], rules of thumb are often created. For instance, it has been established that aging of GFRP in seawater at 50°C is 12 times as severe as that in seawater at 23°C. However, there is no real evidence for the accuracy of this rule. [Silva, 2007]

An attempt to simulate the long-term behaviour of epoxy adhesives outdoor exposed up to three years by accelerated weathering procedures was recently presented. [Lettieri et al., 2006] In order to reach complete curing at the test temperature, samples of three different commercial two-components epoxy adhesives were cured at ambient temperature for 4.5 months and then subjected to accelerated aging. Other samples were outdoor exposed, without any protection from the environmental agents. The accelerate aging conditions used were according to the procedure C of the ASTM standard [ASTM D 756-93, 1993], i.e.: exposure in a climatic chamber at 70±2°C and 75±5% R.H. for different time spans, up to 34 days. This procedure simulates natural weathering conditions in an accelerate manner. High temperature and high moisture levels are the only agents taken into account. Radiation effects are neglected. A similar practice was, in fact, reported as leading to thermo-hygrometric conditions comparable, and close, to those obtained after two years natural aging in the western Mediterranean basin. [Vauthier, 1998] The monitoring of aging, in terms of variations in weight, glass transition temperature, flexural Young modulus, strength and strain at break, measured after both natural aging and the accelerated procedure, produced quite different results. In particular, the fluctuations in Tg of samples outdoor exposed were always lower than the changes measured on the samples subjected to the accelerated aging. Similarly, all the flexural characteristics of the three resins were found to be more strongly affected by the thermo-hygrometric treatment. In conclusion, it was not possible to find any time-correlation between the effects of the accelerated aging and those of natural weathering.

It can be concluded that this kind of investigations cannot be directly correlated with the long-term reliability of the materials under study. The most that can be inferred from these tests is that materials which perform better in the tests will perform better also under environmental conditions.

A rationale prediction through accelerating procedures should include a broad range of service environments, carefully chosen in order to give an high accelerating factor, but still realistic. The conditions selected, in fact, must produce degradation mechanisms similar to those taking place in a material exposed to natural environments. Finally, for each material, a correlation between the results obtained under natural and artificial weathering conditions must be precisely developed. This latter aspect is the hardest to be solved, since it would require a huge number of carefully selected procedures (based on both natural and artificial exposure) performed on a wide variety of materials and their possible combinations (i.e. adhesive resins, fibers, composites).

CONCLUSION

The use of composite materials, based on thermosetting polymers reinforced with fibers (FRP systems), is receiving widespread attention in the construction industry, due to the undoubted advantages of these materials over traditional types.

Even though it is generally accepted that FRP composites, particularly those employing carbon fibers, have a high durability, their long-term performance still needs to be demonstrated. For all the materials the exposure to outdoors conditions leads to a general degradation of properties. In the case of polymer composite, their durability depends equally on the selection of base materials, processing techniques and conditions, loading history and environmental exposure.

Thermosetting resins used as matrix of the composite and adhesives are the most vulnerable components, since they are in an unstable thermodynamic state, mainly because their glass transition temperature is close to ambient temperature. If cross-linking of the resin is not been completed in the initial stages, due to the slowing down of the reaction when cure is conducted at room temperature, a thermal treatment above its Tg would be required to complete the curing process. The exposure to a temperature slightly higher than the ambient temperature, therefore, causes an increase in Tg and the strengthening and stiffening of the system, as a consequence of the increase in cross-linking density.

Absorption of water, both in liquid or vapor form, produces a decrease in the initial Tg of the resin, which affects the mechanical properties. The decrease in Tg, however, enables the post-curing process takes place at lower environmental temperatures.

Another process can take place in cold-cured thermosetting resins, known as Physical Aging. Physical aging occurs in all amorphous polymers at temperatures below the glass transition temperature (Tg). [Struik, 1978] Even though all large-scale molecular motions stop below the Tg, polymer glasses continue to proceed toward thermodynamic equilibrium through small-scale conformational changes. This leads to a reduction in the polymer's free volume over time, i.e. in a "densification", with a consequent modification of all temperature-dependent properties, e.g. lowering of the creep compliance, embrittlement of the aged polymer (stiffening due to aging), reduction in ultimate elongation, and increase in yield strength. [Morgan, 1979; Kong, 1981; Berens and Hodge, 1982; Hodge, 1983; Nichols et al., 1990; Cook et al., 1999]

Structural relaxations in the glassy state are very slow. The relaxation time at 30°C below the glass transition temperature is about 10^{+10} minutes, decreasing to about 1 minute at temperatures close to Tg and to about 10^{-10} minutes at 30°C above the Tg. [Nichols et al., 1990; Clough and Gillen, 1991] Those latter are the case of thermosetting resins used as matrix for composites employed in civil engineering applications, when the external temperature can easily approach and even exceed the Tg of the resin.

However, physical aging is a thermo-reversible phenomenon, which can be erased by heating the polymer above its glass transition temperature. [Struik, 1978; Kong, 1986; Nichols et al., 1990; Montserrat, 1992]

When the service temperature exceeds the Tg of the aged adhesive resins, which may even be reduced due to plasticization effects of water, a "rejuvenation" of the resin takes place, recovering the initial properties. Cold-cured thermosetting adhesives exposed to natural weather, therefore, are constantly subjected to aging and rejuvenation processes that take place in non-isothermal conditions depending on the actual meteorological weather. [Frigione et al., 2001, a)]

In conclusion, the cold-cured resins, used either as matrices and adhesives for FRP composites, are constantly subjected to variations in their thermal (Tg) and mechanical properties during any outdoor exposures. These variations can be reflected also on their adhesive behavior (with fibers as well as with concrete or stone substrates) causing displacements, debonding of fibers and the creation of internal and/or interfacial stresses.

Analyzing in detail the effects of the most common external agents on the properties of cold-cured resins, the following indications can be obtained.

Apart from the mentioned effects, an increase in temperature can be particularly dangerous for cold-cured thermosetting matrices, since their glass transition temperature can be approached, causing a substantial drop in mechanical and adhesive properties. It has generally been found, however, that resins can recover almost completely their properties when the service temperature is lower than their glass transition temperature, i.e. the thermal effects are largely reversible. On the other hand, when resins are not fully cured at ambient temperature and are exposed to temperature higher than their Tg, post-curing reactions can take place, producing a further increase in the Tg of the resin and an enhancement of their mechanical properties.

The effects of water, both in liquid or vapor state, on cold-cured resins include a) plasticization, which lowers Tg and affects the mechanical properties, b) additional curing promoted by the reduction in Tg, which causes a new increase in Tg and the strengthening of the system, c) degradation reactions at long immersion time, with severe reductions in the performance of the system. The presence of salts or chemicals in water do not affect to a great extent the properties of resins relative to immersions in pure water.

Referring to the bond properties of cold-cured adhesives with different substrates, the presence of a wet ambient generally leads to reductions in the adhesion strength. Similarly, repeated freeze-thaw cycles can cause deterioration of the bond strength of cold-cured adhesives, which is mainly attributed to the different coefficients of thermal expansion of the adhesive and of the constituents of the joint.

Combined cycles, using different conditions of temperature, relative humidity, or immersion in different water solutions, are often employed to simulate actual environmental conditions, since under normal service conditions all these environmental agents act simultaneously. When cold-cured resins are exposed to more than one environmental agent, on the other hand, there can be a combined and possibly synergistic effect on the properties of the systems. The change in properties after such combined cycles, however, has not already proved do be easy to correlate with the modifying cause.

Exposure to natural conditions generally causes limited changes of characteristics of cold-cured resins; these changes, moreover, are related to the meteorological climate encountered by the system. The changes in the glass transition temperature (Tg), mechanical and adhesive properties are mainly influenced by the presence of humidity and fluctuations in temperatures.

The scarce number of durability studies specifically focused on polyester and vinyl ester resins cured at moderate temperatures does not allow an accurate evaluation to be made of their long-term properties when exposed to different harsh environments.

Accelerated tests, finally, have the advantage of being fast and repeatable. However, if their only purpose is the prediction of the long-term behavior of a weathered material, this would be not feasible since an exact relationship between the real-time tested performance and that obtained from accelerated testing cannot be easily established. On the other hand, they can certainly provide a reliable guide to the ultimate characteristics of the materials involved.

REFERENCES

Aiello, MA; Frigione, M; Acierno, D. *ASCE J. Mater. Civil. Eng.*, 2002, 14, 185-189.

Aiello, MA; Frigione, M; Leone, M; Aniskevich, AN; Plushchik, OA. *24th Int. SAMPE Europe Conf.* Paris, France, 2003, 239-249.

Aiello, MA; Sciolti, MS. *3rd Int. Conf. Composites in Constructions, CCC* 2005, Lyon, France, 501-510.

Albrecht, P; Mecklenburg, MF; Evans, BM. *Interim Report FHWA/RD*; Federal Highway Administration, Contract DTFH61-84-R-0027, *Washington DC, USA*, 1985.

Antoon, MK; Koenig, JL. *J. Polym. Sci. – Part B Polym. Phys.*, 1981, 19, 197-212.

Apicella, A; Migliaresi, C; Nicodemo, L; Nicolais, L; Iaccarino, L; Roccotelli, S. *Compos*, 1982, 13, 404-410.

Apicella, A; Migliaresi, C; Nicolais, L; Iaccarino, L; Roccotelli, S. *Compos*, 1983, 14, 387-392.

ASTM D 756-93 *Determination of weight and shape changes of plastics under accelerated service conditions*, 1993.

Bank, LC; Gentry, TR; Thompson, BP; Russell JS. *Constr. Build. Mater*, 2003, 17, 405-437.

Bellenger, V; Le Huy, M; Paris, M; Verdu, J. *3rd European Polymer Federation Symp. on Polymeric Materials, EPF* 1990, Sorrento, Italy, 206-207.

Berens, AR; Hodge, IM. *Macromolecules*, 1982, 15, 756-761.

Bisby, LA; Green, MF. *ACI Struct. J.* 2002, 99, 215-223.

Brewis, DM; Comyn, J; Shalash, RJA. *Int. J. Adhes. Adhes*, 1982, 2, 215-222.

Brewis, DM; Comyn, J; Shalash, RJA. *Polymer*, 1983, 24, 67-70.

Chin, JW; Nguyen, T; Aouadi, K. *J. Appl. Polym. Sci.*, 1999, 71, 483-492.

Chu, W; Wu, L; Karbhari, VM. *Compos. Struct*, 2004, 66, 367-376.

Clough, RL; Gillen, KT. In *Irradiation Effects on Polymers*; Clegg, DW; Colleyer, AA; Eds; Elsevier Applied Science: *London, U.K.*, 1991, Chap. 4.

Colombi, P; Fanesi, E; Fava, G; Poggi, C. *3rd Int. Conf. Composites in Constructions, CCC, Lyon, France*, 2005, 291-298.

Comyn, J. In *Engineering Materials Handbook, Adhesives and Sealants*; ASM International: *USA*, 1990, Vol. 3, Section 7, pp 616-621.

Cook, WD; Mehrabi, M; Edward, GH. *Polymer*, 1999, 40, 1209-1218.

Davalos, JF; Boyajian, DM; Kodkani, S; Ray, I; Qiao, P. *1st Int. Conf. on Innovative Materials and Technologies for Construction and Restoration, IMTCR, Lecce, Italy*, 2004, Vol. 2, 10-19.

De'Neve, B; Shanahan, MER. *Polymer*, 1993, 34, 5099-5105.

Di Tommaso, A; Foraboschi, P. *L'Edilizia*, De Lettera Ed. 1995, N. 9/10, 86-116.

Ellis, B. *Chemistry and Technology of Epoxy Resins*; Ellis, B; Ed; Blackie Academic and Professional: *London, U.K.*, 1993, Chap. 3, pp 72-116.

Franke, L. *Int. Symp. Adhesion between Polymers and Concrete*, Paris, France, 1986, 461-473.

Frigione, M; Naddeo, C; Acierno, D. *Mater. Eng.*, 2000, 11, 59-80.

Frigione, M; Naddeo, C; Acierno, D. *J. Polym. Eng.*, 2001a, 21, 23-51.

Frigione, M; Naddeo, C; Acierno, D. *J. Polym. Eng.*, 2001b, 21, 349-367.

Frigione, M; Naddeo, C; Acierno, D. *Mater. Eng.*, 2001c, 12, 443-454.

Frigione, M; Aiello, MA; Leone, M. *11th Int. Cong. on Polymers in Concrete, ICPIC* 2004a, *Berlin, Germany*, 495-502.

Frigione, M; Lettieri, M; Mecchi, AM. *1st Int. Conf. on Innovative Materials and Technologies for Construction and Restoration, IMTCR*, Lecce, Italy, 2004b, Vol. 1, 375-384.

Frigione, M; Aiello, MA; Naddeo, C. *Constr. Build. Mater.* 2006a, 20, 957-970.

Frigione, M; Sciolti, SM. *Int. Symp. Polymers in Concrete, ISPIC, Guimarães, Portugal*, 2006b, 297-304.

Grace, NF; Dalal, V; Ibrahim, E. *Conf. on Structural Composites for Infrastructure Applications, MESC-3, Aswan, Egypt*, 2002, 1-10.

Grace, NF; Grace, M. *3rd International Conference Composites in Constructions, CCC, Lyon, France*, 2005, 391-396.

Green, MF; Bisby, LA; Beaudoin, Y; Labossière, P. *1st Int. Conf. on Durability of FRP Composites for Construction, CDCC, Sherbrooke, Canada*, 1998, 179-190.

Gupta, VB; Drzal, LT; Lee, CYC; Rich, MJ. *Polym. Eng. Sci.*, 1985, 25, 812-823.

Hodge, IM. *Macromolecules*, 1983, 16, 898-902.

Hunston, D; Juska, T; Karbhari, VM; Morgan, R; Williams, C. In *Gap Analysis for Durability of Fiber Reinforced Polymer Composites in Civil Infrastructure*, ASCE Publications: *Reston, VA, USA*, 2001, Vol. 7.

Ishai, O. *Polym. Eng. Sci.*, 1975, 15, 491-499.

Ishida, H; Koenig, JL. *Polym. Eng. Sci.*, 1978, 18, 128-145.

JSCE Research Committee on Continuous Fiber Reinforcing Materials In *Concrete Engineering Series, 3*, JSCE, 1993.

Kahraman, R; Al-Harthi, M. *Int. J. Adhes. Adhes*, 2005, 25, 337-341.

Kajorncheappunngam, S; Gupta, RK; GangaRao, HVS. *ASCE J. Compos. Construct,* 2002, 6, 61-69.

Karbhari, VM. *1st Int. Conf. of Advanced Polymer Composites for Structural Applications in Construction, Southampton University, UK*, 2002, 33-43.

Karbhari, VM. *Reinf. Plast*, 2005, January, 14-25.

Karbhari, VM; Chin, JW; Hunston, D; Benmokrane, B; Juska, T; Morgan, R; Lesko, JJ; Sorathia, U; Reynaud, D. *ASCE J. Compos. Construct*, 2003, 7, 238-247.

Kollek, H. In: *Engineering Materials Handbook, Adhesives and Sealants*; ASM International: USA, Section, 1990, 7, pp 656-662.

Kong, ESW. *J. Appl. Phys.*, 1981, 52, 5921-5925.
Kong, ESW. *Advances in Polymer Science 80*; Springer-Verlag Berlin Heidelberg; Ed; *Berlin, GERMANY*, 1986, pp 125-171.
M. Ladner, *EMPA Bericht nr 37846/4*; *Zürich, Switzerland*, 1983.
Laurenzi Tabasso, M. *Construction and Rehabilitation, Colloque Europeen* Lyon, *France*, 1992, 13-23.
Lesko, JJ; Hayes, MD; Schniepp, TJ; Case, SW. *Int. Workshop Composites in Construction: a Reality Capri, Italy*, 2001, 110-119.
Lettieri, M; Frigione, M; Stefanelli, M. *Int. Symp. Polymers in Concrete, ISPIC Guimarães, Portugal*, 2006, 285-295.
Li, C; Dickie, RA; Morman, KN. *Polym. Eng. Sci.*, 1990, 30, 249-255.
Liew, YS; Tan, KH. *6th Int. Symp. on FRP Reinforcement for Concrete Structures, FRPRCS-6* Singapore, *Singapore*, 2003, 769-778.
Littmann, K. *2nd Int. RILEM Symp. on Adhesion between Polymers and Concrete Dresden, Germany*, 1999, 193-203.
Marshall, JM; Marshall, GP; Pinzelli, RF. *Polym. Compos.*, 1982, 13(3), 131-137.
Mascia, L; Allavena, *A. Reinforced Plastics Congress Brighton*, U.K, 1976, 175-182.
May, CA. In: *ASM Engineered Materials Reference Book*, ASM International: *USA*, 1989, pp 66-77.
Mays, GC; Hutchinson, AR. *Adhesives in Civil Engineering*; Cambridge University Press: *Cambridge, U.K*, 1992.
Meyer, F; Sanz, G; Eceiza, A; Mondragon, I; Mijovic, J. *Polymer*, 1995, 36, 1407-1414.
Mirmiran, A; Shahawy, M. *ASCE J. Struct. Eng.*, 1997, 123, 583-590.
Montserrat, S. *J. Appl. Polym. Sci.*, 1992, 44, 545-554.
Morgan, RJ. *J. Appl. Polym. Sci.*, 1979, 23, 2711-2717.
Mukhopadhyaya, P; Swamy, RN; Lynsdale, CJ. *Constr. Build. Mater*, 1998, 12, 427-446.
Nelson, W. *Accelerated testing: statistical models, test plans and data analyses*; John Wiley and Sons: New York , NY, USA, 1990.
Nichols, ME; Wang, SS; Geil, PH. *J. Macromol. Sci. – Phys.*, B29, 1990, 4, 303-336.
Norris, T; Saadatmanesh, H; Ehsani, MR. *ASCE J. Struct. Eng.*, 1997, 123, 903-911.
Pilakoutas, K; Hafeez, S; Dritsos, S. *Mater. Struct*, 1994, 27, 527-531.
Popineau, S; Shanahan, MER. *Int. J. Adhes. Adhes*, 2006, 26, 363-370.
Renton, WJ. *U.S. Govt. Report AFML-TR-78-127*, 1, AD-A065 500, 2, AD-A065 521, 1978.
Riffle, JS; Lesko, JJ; Puckett, PM. 2nd Int. Conf. on *Composites in Infrastructures Tucson, AZ, U.S.A*, 1998, Vol. 1, 23-34.
Rinaldi, G; Maura, G. *Polym. Int.*, 1993, 31, 339-345.
Rivera, J; Karbhari, VM. *Compos. Part B Eng.*, 2002, 33, 17-24.
Riviere, J. *Construction and Rehabilitation, Colloque Europeen* Lyon, *France*, 1992, 7-16.
Romanko, J; Knauss, WG. *U.S. Govt. Report AFWAL-TR-80-4037*, 1, AD-A087 280, 2, AD-A087 360, 1980.
Sagi-Mana, D; Narkis, M; Siegmann, A; Joseph, R; Dodiuk, H. *J. Appl. Polym. Sci.*, 1998, 69, 2229-2234.
Sen, R; Shahawy, M; Mullins, G; Spain, J. *ACI Struct, J.* 1999, 96, 906-914.
Shaw, SJ. In *Chemistry and Technology of Epoxy Resins*; Ellis B; Ed; Blackie Academic and Professional: *London, U.K*, 1994, pp 206-255.
Sookay, NK; von Klemperer, CJ; Verijenko, VE. *Compos. Struct*, 2003, 62, 429-433.

Struik, LCE. *Physical Aging in Amorphous Polymers and Other Materials*; Elsevier: New York, *NY, USA*, 1978.

Tadeu, AJB; Branco, FJFG. *ASCE J. Mater. Civil Eng.*, 2000, 12, 74-80.

Tan, KH; Liew, YS. *Sci. Eng. Compos. Mater*, 2005, 12, 219-228.

Uomoto, T. *6th Int. Symp. on FRP Reinforcement for Concrete Structures, FRPRCS-6*, Singapore, *Singapore*, 2003, 37-50.

Van Gemert, D; Brosens, K; Ahmed, O. *10th International Congress on Polymers in Concrete, ICPIC* 2001, Honolulu, Hawaii, U.S.A., Paper 50.

Van Gemert, D; Vanden Bosch, M. *RILEM Int. Symp. on Adhesion between Polymers and Concrete Paris, France*, 1986, 518-527.

Vauthier, E; Abry, JC; Bailliez, T; Chatetominois, A. *Compos. Sci. Tech.*, 1998, 58, 687-692.

Vilkner, G; Meyer, C. *11th Int. Cong. on Polymers in Concrete, ICPIC* Berlin, *Germany*, 2004, 477-484.

White, JR; Turnbull, A. *J. Mater. Sci.*, 1994, 29, 584-613.

Wolff, L; Hailu, K; Raupach, M. *Int. Symp. Polymers in Concrete, ISPIC* Guimarães, *Portugal*, 2006, 213-223.

Wypych, G. *Handbook of Material Weathering*; ChemTec Publishing: Toronto, CANADA, 1995.

Zhang, S; Karbhari, VM. *14th Tech. Conf. of the American Society of Composites Fairborn, OH, U.S.A.*, 1999, 12-19.

Zhang, S; Karbhari, VM; Ye, L; Mai, YW. *J. Reinforced Plast. Compos*, 2000, 19, 704-731.

Zhou, J; Lucas, JP. *Polymer*, 1999a, 40, 5505-5512.

Zhou, J; Lucas, JP. *Polymer*, 1999b, 40, 5513-5522.

In: Adhesive Properties in Nanomaterials, Composites ... ISBN: 978-1-61209-268-3
Editors: K. A. Wilkinson and D. A. Ordonez

Chapter 8

REPLACING OF SYNTHETIC ADHESIVES WITH NATURAL ADHESIVES[†]

Md. Moniruzzaman Khan and M. Rafiqul Islam [*]
Department of Civil and Resource Engineering, Dalhousie University
D510-1360 Barrington Street, Halifax, NS B3J 2X4, Canada

ABSTRACT

Adhesives or bonding agents surround all living beings in Nature and in their daily lives. Adhesives or bonding agents are used in a variety of industries: construction, packaging, furniture, automotive, appliance, textile, aircraft, and many others. However, most of them are toxic for human beings due to the presence of harmful synthetic additives. The long-term exposure to these toxic substances can cause a range of ailments, including cancer, asthma, and Alzheimer disease. The ultimate solution of this trend lies in the emulation of Nature following the true pathway of natural process. In this paper, a number of adhesives have been formulated from natural additives, which can lead us towards sustainable lifestyles. Experiments show that the product exhibit durability and strength comparable to commercially available products, while the required concentration of the adhesive is low. Further experiments indicate that alternative adhesives that are of organic origin can eliminate the use of synthetic adhesives entirely. Similarly, as a natural alternative to Plaster of Paris is proposed. This product has features very similar to the commercial Plaster of Paris. This product and others can become useful for people with chemical sensitivity, particularly the children and the elderly. This paper describes in detail the specific recommendations on natural adhesives, their strengths and durability as a function of temperature.

Keywords: natural adhesive; synthetic adhesive; book binding glue; particle board.

[†] A version of this chapter also appears in *Journal of Characterization and Development of Novel Materials*, Volume 1, Number 2, published by Nova Science Publishers, Inc. It was submitted for appropriate modifications in an effort to encourage wider dissemination of research.

[*] Corresponding author: E-mail: rafiqul.islam@dal.ca

1. Introduction

Adhesive is a substance capable of holding at least two surfaces together by means of surface attachment. The use of adhesives in human life has a long history, which dates back to ancient times. Adhesives were first used thousands of years ago. Humans have originally opted for natural adhesives. Early hunters may have seen improvement in their aim by bonding feathers to arrows with beeswax, a primitive, yet natural, form of adhesive (Petrie, 2000). Most adhesives evolved from vegetable, animal, or mineral substances. However, as human technology advanced and chemical industries were available on a commercial scale, the use of man-made processes accelerated. Soon after the industrial revolution, artificial and toxic additives rapidly began to replace previous natural adhesives. A brief historical development of adhesive replacement from natural to synthetic polymeric adhesives has been presented in Table 1.

Table 1. Historical development of adhesives and sealants (Petrie, 2000).

Approximate decade of commercial Availability	Adhesive or sealant
Pre 1910	Glue from animal bones, Fish glue, Vegetable adhesives
1910	Phenol-formaldehyde, Casein glues
1920	Cellulose ester, Alkyd resin, Cyclized rubber in adhesives, Polychloroprene (Neoprene), Soybean adhesives
1930	Urea-formaldehyde, Pressure sensitive tapes, Phenolic resin adhesive films, Polyvinyl acetate wood glues
1940	Nitrile-phenolic, Chlorinated rubber, Melamine formaldehyde, Vinyl-phenolic, Acrylic, Polyurethanes
1950	Epoxies, Cyanoacrylates, Anaerobics, Epoxy alloys
1960	Polyimide, Polybenzimidazole, Polyquinoxaline
1970	Second-generation acrylic, Acrylic pressure sensitive, Structural polyurethanes
1980	Tougheners for thermoset resins, Waterborne epoxies, Waterborne contact adhesives, Formable and foamed hot melts
1990	Polyurethane modified epoxy, Curable hot melts, UV and light cure systems

It is observed that the age of modern (and artificial or synthetic) adhesives began about 1910 with the development of phenol formaldehyde adhesives for the plywood industry. After that period, significant incremental developments took place only for synthetic adhesives. The natural adhesives were over looked almost in every decade after 1910. The demand for adhesives in the U.S. was forecasted to rise to 14 billion pounds in the year 2001, with market value reaching $9 billion (Petrie, 2000). Today adhesives are used within virtually every business and industry. Figure 1 shows the use of adhesive in different areas of daily modern life in United States.

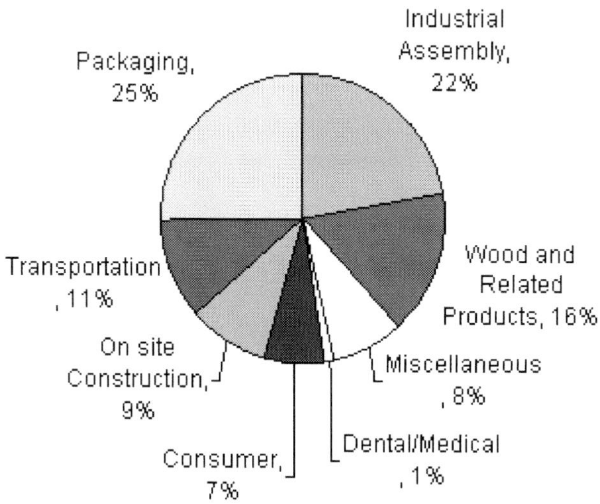

Figure 1. Adhesive market in USA in 1995 (Modified from Prane, 1996).

There are many types of adhesives, massively used for both structural and non-structural applications. Table 2 shows a list of some common types of adhesives most of which are synthetic in nature.

Table 2. List of adhesive of different kinds (Modified from Petrie, 2000).

Thermoplastic resins	Thermosetting resins	Elastomeric resins	Other materials used as adhesives
Polyvinyl acetate	Epoxy	Natural rubber	Agricultural glues
Polyvinyl alcohol	Polyimide	Asphalt	Animal glues
Thermoplastic	Polybenzimidazole	Butyl rubber	Sodium silicate
Cellulosic resins	Bismaleimide	Chloroprene Acrylonitrile	Phosphate cements
Polyamide	Acrylic	Polyisobutlyene	Litharge cement
Polysulfone		Polysulfide	Sulfur cement
Phenoxy		Silicone	
Acrylic			

2. Toxic and Hazardous Properties of Adhesives

Most of the synthetic adhesives are found toxic and hazardous for human beings. The toxicity and hazardous properties of adhesives are determined by analysis of the followings factors (Sabater et al., 2001):

1. Flash point;
2. Corrosivity;
3. Reactivity (explosive, stable character, formation of toxic gases in contact with water);
4. Carcinogenic or probable carcinogenic substances;
5. Mutagenic or teratogenic substances;
6. Toxicity and
7. Ecotoxicity.

However, carcinogenicity, mutagenicity, toxicity and ecotoxicity are the most important factors to be addressed to determine the sustainability of an adhesive in the long term.

2.1. Toxicity from Existing Wood Panels

Whatever the raw materials or wood size, every wood product uses adhesive resins that fall under the following main categories (Lambuth, 1977):

- Urea-formaldehyde (for internal use and with greater resistance to mechanical stress)
- Phenol-formaldehyde (external use and greater resistance to humidity)

Formaldehyde is a colorless and odorous gas. It is reported that when present in the air at levels above 0.1 ppm, it can cause watery eyes, burning sensations in the eyes, nose and throat, nausea, coughing, chest tightness, wheezing, skin rashes, and allergic reactions (CPSC, 2005). Formaldehyde, acetaldehyde, and acrolein have been declared toxic under the Canadian Environmental Protection Act (LaFleur et al., 2004). It is found that the wood panel and building materials from wood panel emits volatile organic components (VOC) which pose threats of discomfort and injury to health. The toxicity of formaldehyde to human health can be divided into the following categories:

- Toxicity of short term exposure
- Toxicity of long term exposure
- Acute toxicity

Short-term inhalation exposure can result in eye, nose, and throat irritation and respiratory symptoms. Chronic formaldehyde exposure can cause menstrual disorders and pregnancy problems in women workers exposed to high levels. Formaldehyde is considered as the possible source of human carcinogen when inhaled or ingested (Popa et al., 2002). The formaldehyde glues in oriented strand board (OSB) and plywood are proven carcinogens and asthma triggers. Persons have developed allergic reactions (allergic skin disease and hives) to formaldehyde through skin contact with solutions of formaldehyde or durable-press clothing containing formaldehyde (CPSC, 2005). Others have developed asthmatic reactions and skin rashes from exposure to formaldehyde.

The emission of formaldehyde into the environment can be spread out and cause air/water pollution in different ways:

2.1.1. Indoor Air Pollution

Indoor air quality problems resulting from the emission of volatile organic compounds (VOCs) have become an issue of increasing concern. Hundreds of VOCs have been detected in the indoor air (Popa et al., 2002 and Baumann et al., 2000). The tendency of saving energy by reducing ventilation rate in the cold countries has deteriorated the indoor quality of air. Emissions from building and furnishing materials, which are frequently constructed from particleboard and medium density fiberboard (MDF), are a potentially important contributor of indoor VOCs.

2.1.2. Air Pollution in the Work Place

During wood panel production, the synthetic glue materials change the ambient air and increase the toxicity of the product (Yakovskaya, 1993). It has been reported that residents in a community adjacent to a particleboard plant raised concerns regarding odors and respiratory symptoms when the winds were from the direction of the plant (Elias et al., 1996). Formaldehyde was primarily released to the environment through the Press/Boiler stack during drying and manufacturing. Stack measurements showed that 445 g/hr of formaldehyde were emitted from the 80 ft. stack. Residential exposures to formaldehyde are quite common as this compound is used in many industrial processes (Elias et al., 1996 and Johansson et al., 2004).

2.1.3. Indirect Air Pollution

The combustion of wood panel made from these toxic glue release significant amount of formaldehyde vapor to the environment and pose toxicity to the environment. In addition, the oxidized products that arise from the oxidation of trace elements (added during chemical processing of the adhesive) remain the most harmful, yet vastly undetected pollutants of the environment. Conventional analysis and toxicity evaluation do not include these chemical products as they are mainly below detection limit. Yet, each of these toxins has profound impact on human health due to their cumulative effects.

2.1.4. Indirect Water Pollution

Chemicals derived from phenolformaldehyde glue play a significant role in the toxicity of wastewater, which is primarily due to wood extractives (Rogers et al., 1979). Here, it should be noted that conventional analysis techniques do not consider the difference between an organic product and a synthetic product. For instance, naturally occurring dioxin is considered to be the same as dioxin released from PVC. Without clear distinction between these two chemicals of starkly diverging pathways, it is not possible to quantify, let alone proposing a remedy to the pollution caused by synthetic products.

It is found that wood panels made with formaldehyde adhesives attribute different types of toxicity with no remedy. Toxicity continues with time and ultimately humans suffer direct or acute injury to health. If natural substitutes to synthetic formaldehyde were to be introduced, this would not be case. Paraphrasing the slogan of a natural health product manufacturer (supplier to Air Canada), one can say, no toxicity testing is needed for natural products. Natural, here, refers to products that are of real origin and pathway (Zatzman and Islam, 2007).

3. Sustainable Technology for Adhesive Preparation

There is a worldwide trend towards the development of adhesives that are environmentally benign. The term, "environmentally benign" indicates that there are no toxic materials released to the air, water or soil when the adhesive is being manufactured, applied, cured, reused or recycled throughout its intermediate and final product lifetime. The ultimate solution of this trend lies in the emulation of Nature following the true pathway of natural process. Khan and Islam (2007) described the characteristic features of nature. The difference between natural and synthetic materials shows that natural materials are very non-linear, complex but shows unlimited adaptability. With the first premise of 'Nature is perfect', any technology that conflicts with natural traits will not be sustainable. That is why, natural adhesives additives are reviving due to its environmentally friendly and sustainable features.

3.1. Natural Additives for Formulating Natural Adhesives

Many adhesives are made from organic polymers. There are also adhesives with inorganic origin. The oldest polymers used for adhesives were of natural origin. Many of these adhesives are used because of their early history and the number of industries that have adopted and continued to use these systems. Often naturally occurring adhesives are thought to be inferior to synthetic polymers because of their lower strength and limited freedom in processing. However, in many applications, such as bonding of paper and wood where the emphasis may be on the adhesive being biodegradable, naturally occurring adhesives find a strong market. In certain applications, they dominate over synthetic adhesives. There are several adhesives that are derived from agricultural sources such as starch, dextrin, soybean, and oleoresins. Many of the most important natural adhesives derive from animal origin such as casein, blood, bone, and hide byproducts and fish.

Even though several natural adhesives are exist today, the development of totally new natural adhesives is not increasing as compared to synthetic adhesives. This is in stark contrast to the pharmaceutical industry that continues to look for nature to derive ingredients for new drugs (Steenhuysen, 2007). However, it is important to replace the synthetic adhesives by natural adhesives. That is why focusing to develop more sustainable adhesives should be one of the purposes of green technologies. In this paper, several new adhesives have been introduced, which are derived from 100% natural sources such as honey, sugar, mangosteen fluid, CaO etc. Honey is a natural and complex food product produced by bees from nectar of plants and also from honeydew. It is a unique sweetening agent that can be used by humans without any farther processing.

Typical honey analysis (Shin and Ustunol, 2005):

- Fructose: 38.5%
- Glucose: 31.5%
- Maltose: 7.2%
- Sucrose: 1.5%
- Variety of oligosaccharides: 4.2% and
- The balance is water and unknown chemicals.

However, the exact analysis is still unknown. It is known that honey is the most complete food and contains over 250 ingredients, most of which are unknown (Wilson, 2004). These ingredients have been mainly ignored because their proportion is relatively small. However, true analysis should include their role, without which honey becomes equal to a mixture of fructose, glucose, etc. and there remains no difference between honey, sugar, saccharine or aspartame. With true analysis, honey stands out as a unique remedy to numerous health problems, some of which are only beginning to surface in the post-renaissance world, after being known for centuries in other civilizations (AFP, 2007). On the human health side, the assertion that honey is simply a mixture of sweetening agents is fatal and in scientific investigation such assertion borderlines on commercial fraud (Zatzman, 2007). Our study considers the fact that honey is not similar to another sweetening agent with 'similar' composition, it is uniquely different because it is totally natural. In this paper, honey was used with CaO in different proportion to obtain different grade of adhesives. CaO was collected from the calcination of mollusks. Because of biological origin and formation, these biomaterials have superior performance to the synthetic or non-natural structure, for both the mechanical strength and the functional properties. The outstanding properties are only visible and attributed to the biomaterials as observed by their well organized structure and strong interfacial interaction between bio-macromolecules and inorganic components.

4. Materials and Methods

4.1. Testing Materials

In this study, natural adhesives were derived from 100% organic and/or naturally derived additives, which were mixed in different ratio. Additives that were used are as follows:

- Honey
- Sugar
- Egg yolk and white
- Mangosteen fluid
- CaO from sea shales

Pure Nova Scotia Honey (unpasteurized) supplied by Cosman and Whidden was used in this study.

Natural Icing Sugar was supplied by Lantic Sugar Ltd., Canada.

Egg yolk and albumen was collected from chicken eggs by separating the egg yolk from clear liquid (albumen) carefully.

Sea shell, the hard outer shell of a mollusk, is actually calcium carbonate which is grown by a biological process. Sea shales (Figure 2) were collected from Point Pleasant Park, Halifax, NS, Canada. The calcination of sea shell produces quick lime (CaO) (Figure 3) which can be done simply by igniting the shell in the atmosphere temperature and pressure. Calcination was performed by Mining Laboratory, Dalhousie University, Canada. It was then sieved to obtain a size less than 140 US mesh opening.

Figure 2. Sea shales collected from point pleasant park, Halifax, NS, Canada.

Figure 3. Quick lime (CaO) after calcination of sea shales.

The Mangosteen is one of the most praised of tropical fruits, whose scientific name is *Garcinia Mangostana*. Mangoeteen fruits (Figure 4) were collected from Pete's Frootique, Halifax, NS, Canada. The fruits were cut into pieces and heated slowly with constant stirring in order to extract the viscous fluid (Figure 5) from it. Finally, the mixtures were filtered to obtain the clear fluid.

Figure 4. Mangtosteen fruit.

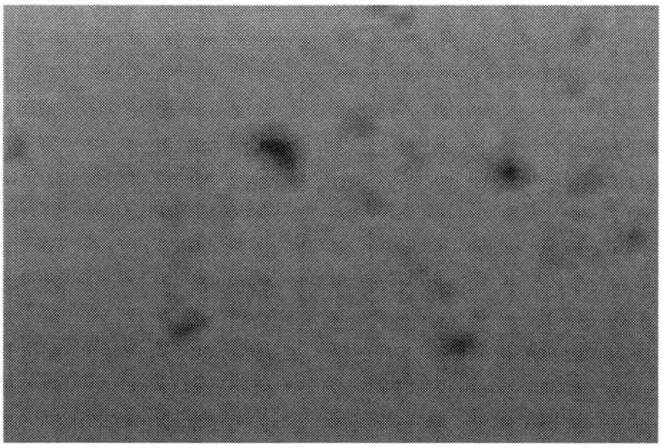

Figure 5. Mangosteen fluid under microscope.

4.2. Formulation of Adhesives

Natural additives were mixed in different proportions to observe the adhesive properties of the combinations. The experimental combinations were as follows:

- Egg (yolk) + CaO
- Egg (White) + CaO
- Egg (yolk) + Honey + CaO
- Egg (yolk) + mangostein fluid + CaO
- Mangostein fluid + Honey + CaO
- Honey + CaO
- Sugar + CaO + Water

4.3. Testing Media

After making the adhesives, the successful adhesives were applied in the following different media to investigate the effectiveness of the adhesive:

Hard Maple blocks were used in selected experiments. These blocks were straight grain and free from defects. The blocks were cut in such a way that the grain direction was parallel to the direction of the loading during test. The dimension of each block was approximately ¾" by 2" by 1¾".

Sawdust of pine wood was used in this series of experiments to prepare a particle board with the adhesive.

Xerox Premium Bright White 98 paper was used in selected experiments. This was bond paper with a basis weight 20 lb and thickness 0.1 mm. 50 sheets of 7.7 cm × 12.7 cm size were stacked together to be used in this experiment.

Nova Scotia sand was used to make sand blocks with natural adhesives. Glasses and ceramics were broken and added by using natural adhesives. Adhesive was applied to cotton cloth to see the applicability of this final product to be used as alternative to Plaster of Paris.

4.4. Testing Method and Standards

ASTM D 898-05 and ASTM D 899-00 methods were used for measuring the adhesive weight per unit area of dry and wet adhesive, respectively.

ASTM D 905-03 method was used for measuring the strength properties of adhesive bonds in shear by compression loading.

4.5. Experimental Equipment

A shearing tool, as shown in Figures 6a and 6b, was used in this study to follow ASTM D 905-03 method for measuring the strength properties of adhesive. This experimental set up was available in the Geomechanical and Petroleum Laboratory of Dalhousie University.

Figure 6a and 6 b. Shearing tools.

4.6. Procedure for Measuring the Strength of Adhesive

Two maple wooden blocks were glued together by using natural adhesive in this study. The blocks were glued together according to Figure 7 following ASTM D 905-03. The glue joint was in the direction of grain direction keeping ¼ in empty from the edge of one block to other block. After gluing, the blocks were given sufficient adhesive drying time.

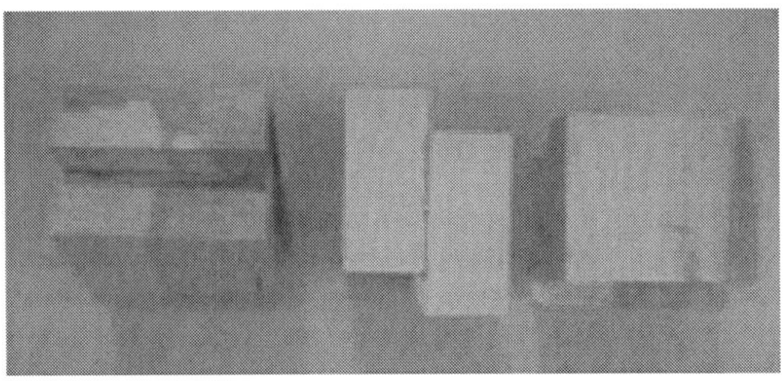

Figure 7. Adhesive application in maple blocks for shear stress measurement.

The blocks were then placed in the shearing tool so that the load may be applied according to Figure 6a and 6b. The shear stress at failure based on the bond line area between two blocks was noted. With time adhesive properties in the glue joint of the maple blocks can be changed. That is why several tests were performed at different times after gluing.

4.7. Results and Discussion

At present, it is practically impossible to get non-toxic glue. Some industrial products have been claimed as 'green'. However, in-depth analysis cannot substantiate this claim. Chhetri and Islam (2007a) demonstrated this in the context of biodiesel and their more recent works also point to similar conclusions regarding other 'green' products (Chhetri and Islam, 2007b). Even though a product (glue, in this particular case) might be produced from organic materials, the use of artificial additives or processes ultimately makes the product toxic and unsustainable. This was substantiated in the recent work of Miralai (2006). As a search for natural glue different natural additives were mixed together to observed the gluing properties. Followings are the results of primary search:

4.7.1. Natural Adhesives

4.7.1.1. Egg (yolk) + CaO

Egg yolks are considered to be important binding agents in many preparations in cooking due to the emulsifying action of lecithin. They were used extensively in the past and are still used in environment-friendly applications of paints, food additives, etc. Egg yolk mainly contains 16.4% protein, 32% lipid and water with a number of minerals and vitamins listed in USDA (2007). Kiosseoglou (2003) speculated the ability of gel formation by egg yolk due to the presence of high lipid molecules. On the other hand, Calcium oxide (CaO) mainly works as a drying agent. In addition, the cations of CaO might take part to increase the adhesive strength. Because of its biological origin, the current sample of CaO contains a number of other trace materials, which also might be related to the binding process.

During initial trials, a small amount of CaO was added to the egg yolk to prepare this glue. However, it was found to dry very quickly. Therefore, it is suggested that the glue be should applied very quickly onto any surface before drying. This is equivalent to binding two chemicals just prior to application, as is the case in the line of products, commercialized as 'super glue'.

The addition of more CaO was not found to be satisfactory. The presence of excess CaO resulted in the formation of solid flakes and reduced bonding strength. This glue was found to be water insoluble. However, the strength of the glue was not acceptable for the range of concentrations studied.

4.7.1.2. Egg (White) + CaO

Egg white is the clear liquid contained within an egg. It is also known as albumen. It contains protein but little or no fat. It is also known as albumen. Egg white albumen (EWA) protein is widely used for various food applications. The EWA, being a protein, contains various functional groups such as –NH2, –COOH, –SH and –OH. Rathna et al. (2004) identified the ability of EWA to form a non-reversible gel at temperatures above 80° C due to covalent disulfide (S–S) bond formation. Dhara and Bhargava (2001) used albumin, the major protein in egg white as a binder for ceramics. That is why it was preferred as a binding agent for adhesive. However after adding CaO with egg white some solid flakes readily formed. No bonding strength was observed.

4.7.1.3. Egg (yolk) + Honey + CaO

Honey is considered to have natural adhesive properties. During the preparation stage, the glue became hot when egg (yolk) was added to honey and CaO, showing exothermic nature of the chemical reaction. As a result, quick drying was observed. A good adhesive property was noticed when it was applied to join two wooden blocks.

4.7.1.4. Egg (yolk) + Mangosteen fluid + CaO

Mangosteen fluid is commonly used in tropical countries as an adhesive in boat building or in manufacturing of fishing nets. It is known for its water resistant properties so that those products do not rot under water through microbial activities. After adding mangosteen fluid with the mixture of egg yolk and CaO, a black color adhesive was formed. It had gluing properties which was found to be water insoluble.

4.7.1.5. Mangosteen fluid + Honey + CaO

Both mangosteen fluid and honey have adhesive properties. Mixing these with CaO increases its adhesive strength. However, it has a long drying time. With a ratio of 1:4:1 (magosteen fluid:honey:CaO), it was found to have a shear stress of 150 psi. Adding mangosteen fluid increases its inclination to water insolubility.

4.7.1.6. Honey + CaO

The combination of honey and CaO showed excellent adhesive properties. This mixture was water soluble. The drying time and adhesive strength varied with the mixing ratio as shown in Table 3.

Table 3. Drying time and bonding strength of honey to CaO ratio on wood surface joint.

Honey: CaO (g/g)	Average Adhesive Weight (g/cm^2)	Average Drying Time (hours)	Average Shear Stress at Bond Failure (psi)
2:1	0.123	1.2	91
4:1	0.086	2.4	143
6:1	0.063	6	134
8:1	0.052	11	83
12:1	0.041	16	50

From Table 3, it is found that with the increase of honey-to-CaO ratio, the adhesive requirements decrease due to an increase in the moisture content. From Figure 8, it is observed that the drying time increases with the increase of honey-to-CaO ratio. However, the shear stress does not follow any proportionality with the mixing ratio. The best strength was found at a ratio of 4:1, immediately after the complete drying of adhesive between the joint of maple wood blocks. The bonding strength of 6:1 honey to CaO also showed good bonding strength. With low honey to CaO ratio (2:1), the adhesive strength was not found satisfactory due to granular nature of the adhesive after drying.

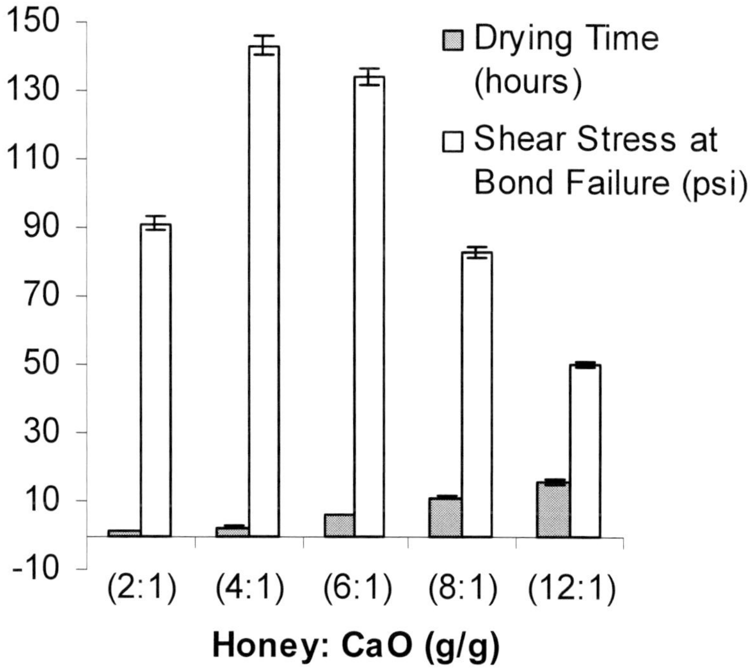

Figure 8. Drying time and bonding strength of honey to CaO ratio on wood surface joint.

4.7.1.7. Sugar + CaO + Water

This combination showed good adhesive strength and a short drying time. The mixing temperature rises to nearly 60° C, which in fact reduces the drying time. The strength of this adhesive varied with the mixture ratio, which has shown in Table 5.

4.7.2. Selection of Adhesives

After considering the adhesive properties and strength of the above experimental results, the blends of Honey + CaO and Sugar + CaO + Water were taken for further experiments trials in order to find their properties with different mixing ratios and the application in different adhesive fields.

4.7.3. Test Results of Adhesive Strength

Honey was mixed with CaO at different ratios and was applied to maple blocks to make a bond (wooden joint) between two blocks following ASTM D 905-03. Table 3 shows the adhesive requirement, bonding drying time and adhesive strength at different ratio of the mixture at a temperature of 24° C.

Another set of experiments was performed following the above experiments to observe the dynamics of the bonding strength between two maple blocks that were joined together with an adhesive. Two mixing ratios (4:1 and 6:1 honey to CaO) were considered in these experiments. The experimental results are as follows:

From Figure 9, it is observed that the bonding strength does not decrease with time; rather it increases, albeit slightly, with time. However, with time the honey to CaO ratio of 6:1 showed a consistency comparable to that of the mixture of the ratio of 4:1.

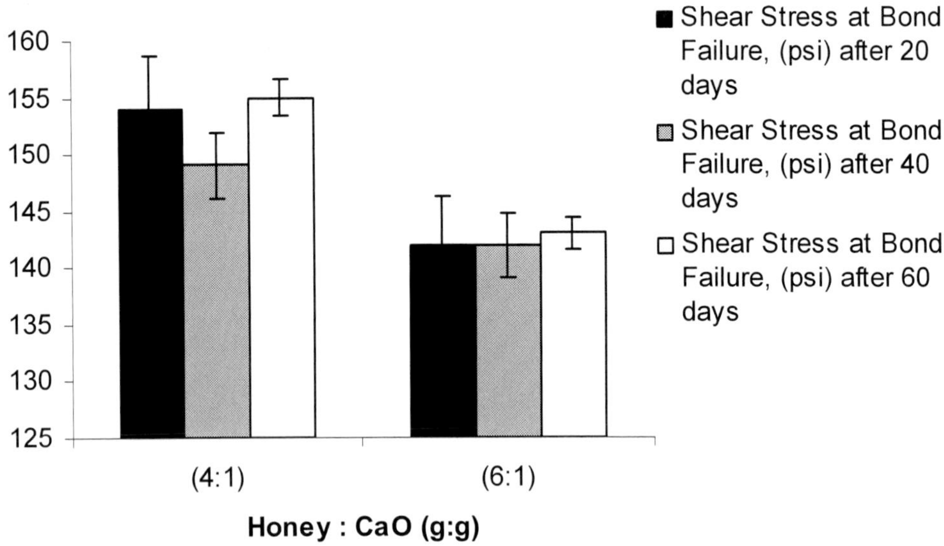

Figure 9. Shear stress at bond failure with the age of adhesive.

According to the above results, Honey to CaO with a ratio of 4:1 can be recommended for general wood joint applications. Finally, another set of experiments were performed at elevated temperatures to observe the effect of heat on adhesive. Only honey to CaO at a ratio of 4:1 was taken into consideration.

These experiments were performed in two different ways. In the first experiment, only honey was heated to a temperature of 130° C and then mixed with CaO. The mixing temperature was 75° C before applying to the wood surface. In the second experiment, the mixture of honey and CaO (4:1) was heated to a higher temperature (160° C) and then applied to wood surface before it cooled. A series of experiments were carried out to find out the optimum temperature for the latter experiments. It was found that the temperature of the mixture above 160° C was unsatisfactory. The mixture starts burning above this temperature. However, it shows very good bonding strength at a temperature of 160° C. That is why the mixture was heated to a temperature of 160° C and then applied to wood surface at a temperature 100° C nearly. Table 4 shows the experimental results at higher temperatures.

Table 4. Adhesive properties at elevated temperature onto wood surface

	Maximum Temperature (°C)	Application Temperature (°C)	Adhesive Weight (g/cm^2)	Drying Time (Hours)	Shear Stress at Bond Failure (psi)
Heated honey with CaO	130	75	0.072	4	150
Heated Glue Mixture	160	100	0.070	1	172

From Table 4, it is found that heated glue mixture perform better. Therefore, among the glue mixture and applications discussed earlier, heated glue mixture of honey to CaO with a mixture ration of 4:1 is the best for the application in the wood surface joints. This does not only increase the bonding strength but also decreases the drying time and material requirements.

The use of this adhesive is recommend for interior applications due to water soluble properties. As this glue is made from 100% natural matters, it is superior to synthetic glue in terms of toxicity.

4.7.4. Application of the Adhesives

4.7.4.1. Book Binding Adhesive

After successful applications on wooden surface of the natural adhesive, a number of experiments were carried out to develop an environment-friendly book binding adhesive. Bookbinding is the process of physically assembling a book from a number of folded or unfolded sheets of paper or other material. The same glue can be used to bind other packages of papers, such as note pad, self-stick notes (e.g. Post-It®), stickers, and numerous others. In this experiment, 50 sheets of Xerox Premium Bright White 98 paper of 7.7cm × 12.7 cm size were assembled for adhesive application to bind them. It is found that honey and CaO with a ratio 6:1 showed satisfactory result for this binding. It took only three hours to dry completely. The binding strength of the adhesive of experimental writing pad was compared with that of thick writing pad supplied by Buffalo-Eastcantra Inc (Figure 10). Air drying was performed, in order to avoid interfering with natural drying process. Fifty sheets from each of writing pads were taken out to measure the strength of the adhesive to bind those sheets together. One single sheet was suspended from the stack of sheets (Figure 11) and then its pulling strength was measured by increasing force applied in the rest of the sheets. The minimum force required to separate a single sheet from the stack of sheet was observed and recorded as its minimum pulling strength.

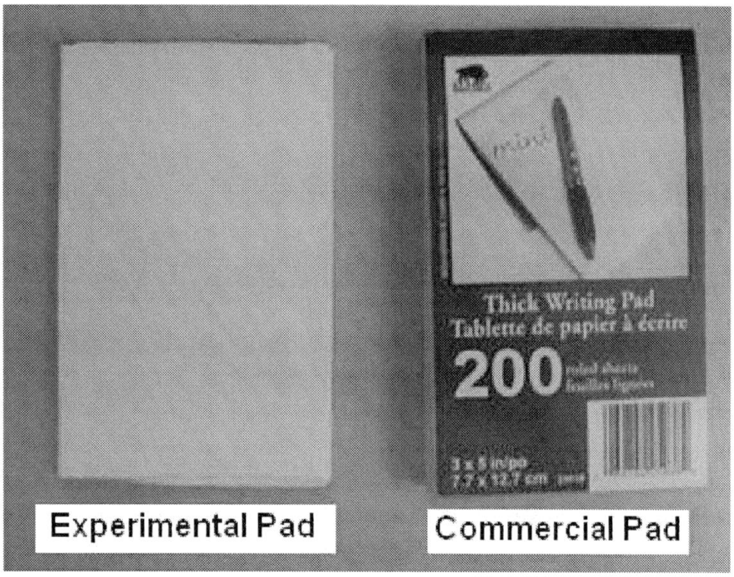

Figure 10. Experimental writing pad and commercial writing pad.

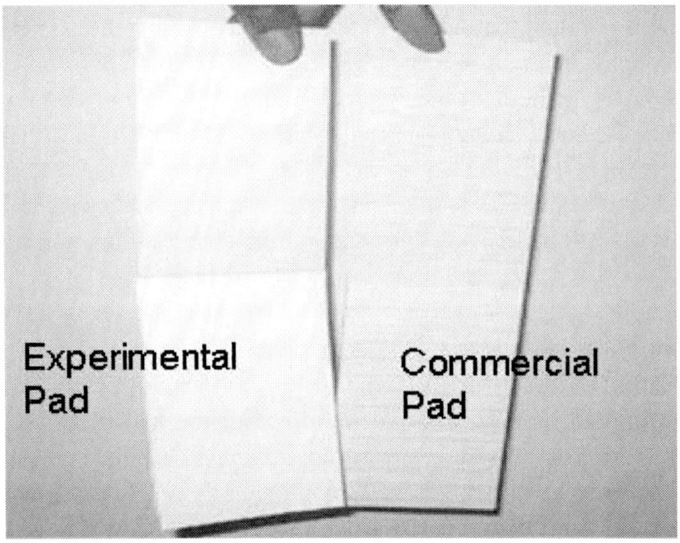

Figure 11. Test of pulling strength of a single sheet in the writing pad.

The experiments were performed several times to reach an average pulling force. Figure 12 shows the average pulling force to separate a single sheet from a writing pad. It is found that the pulling strength does not vary significantly between the two types of writing pads. However, as a non-toxic adhesive the experimental writing pad can get higher acceptancy, particularly for applications involving children, elderly people, and people of chemical sensitivity.

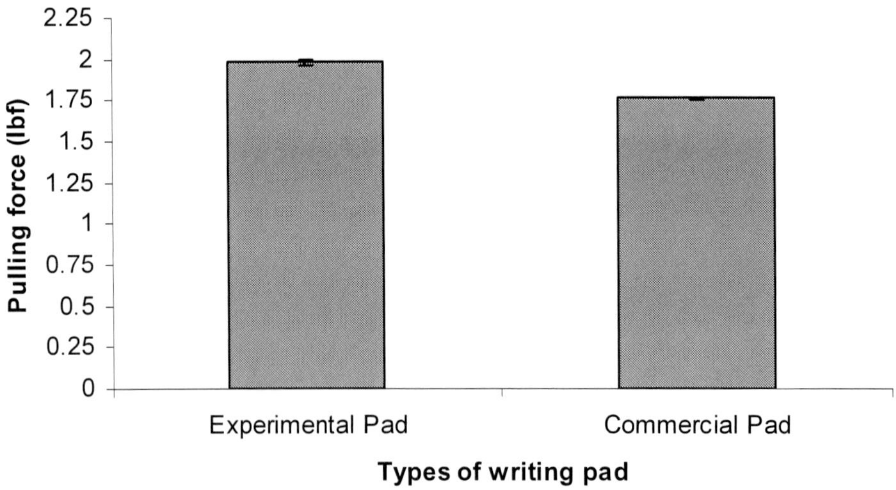

Figure 12. Pulling force of a single sheet to separate from writing pad.

4.7.4.2. Adhesive for Children and People of Special Needs

Adhesive made up of honey with CaO at a ratio of 12:1 showed very sticky. This adhesive can be used by children as a non-toxic adhesive for making paper-based toys. This line of product can also be useful for children and adult who are sensitive to environmental pollution or are susceptible to allergies to synthetic chemicals.

4.7.4.3. Adhesive for Postal Stamps and Envelops

This adhesive can be used in postal stamps and envelops for making a sticky surface that does not dry easily. It is found that honey and CaO with a ratio of 12:1 when heated to a temperature of 120° C showed excellent adhesive properties to be used for making sticky surface on paper. The adhesive was applied at a temperature of 80° C. It took only 8 minutes to dry. Later, the gluing surface was moistened to stick the paper surface to another paper. Usually saliva is used to moistening the glued surface to make it sticky. However, this practice is threatened to public health if the glue is toxic. For the adhesive under this study, it is claimed to be non-toxic due to the use of natural, food grade matters.

4.7.4.4. Adhesive for Ceramics

To join broken ceramics and glasses with adhesives, some experiments were carried out. Two types of adhesives were prepared for this purpose. In the first experiment, sugar and CaO were mixed with water to prepare glue at different mixing ratios and used as an adhesive for joining the broken ceramics (Figure 13). In the second experiment, honey and CaO were mixed at different ratios to observe the adhesive strength for joining the broken pieces of ceramics (Figure 14). Table 5 shows the experimental results of adhesive in the application to join broken ceramics.

Figure 13. Fixing of broken ceramics with sugar and CaO mixture with water.

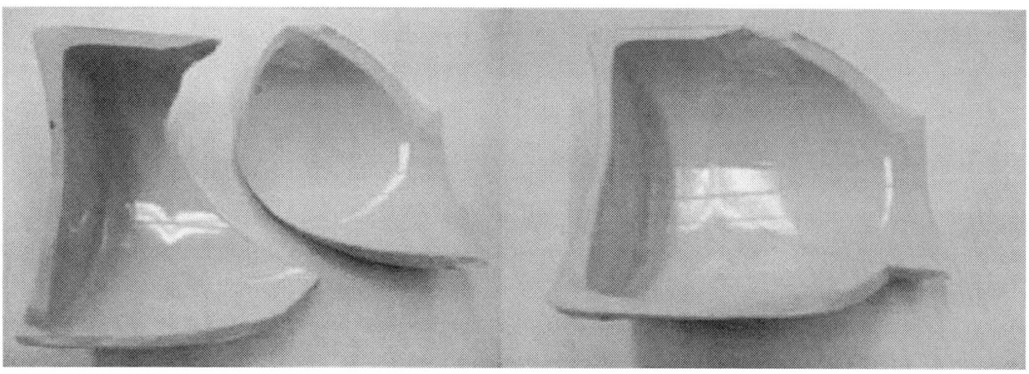

Figure 14. Fixing of broken ceramics with honey and CaO mixture.

Table 5. Adhesive preparation for using to join ceramics pieces.

	Mixing Ratio	Average Drying Time	Average Tensile Adhesive Strength (psi)
Icing Sugar: CaO (water added)	1:3	21 minutes	116
	1:2	14 minutes	217
	1:1	10 minutes	341
	2:1	6 minutes	226
	3:1	3 minutes	165
Honey : CaO	1:1	20 minutes	86
	2:1	1 hour	362
	4:1	3 hours	178
	6:1	8 hours	68

From Figure 14, it is found that icing sugar and CaO at a ratio of 1:1 showed better performance in terms of bonding strength. Adding more sugar reduces the drying time, however, does not increase the strength due to granular nature of the adhesive. On the other hand, adding more CaO the adhesive loses its sticky nature. Due to quick drying time, it is necessary to apply specimens very quickly. For honey glue, honey and CaO at a ratio of 2:1 showed superior performance. Even though it took longer time to dry, it showed better strength as compared to the sugar and CaO mixture. The adhesive was yellow in color. For applications that need the color to be white, icing sugar and calcium oxide can be used.

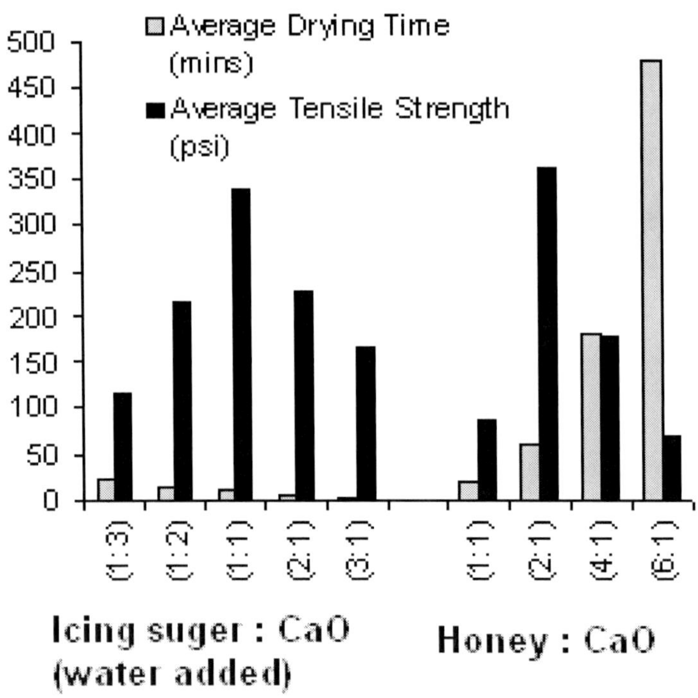

Figure 14. Drying time and tensile strength of adhesive at different ratio.

4.7.4.5. Alternative to Plaster of Paris

Plaster of Paris has been found very useful for its binding properties. It has long been used for the construction of plaster board, insulation board, slabs, mold, tiles etc. Plaster of Paris ($CaSO_4$. ½ H_2O) is derived from gypsum ($CaSO_4$, 2 H_2O), a sedimentary rock. In this study, an alternative to Plaster of Paris was sought. If quick lime (CaO) is mixed with water, it forms calcium hydroxide ($Ca(OH)_2$) known as slaked lime. When the slaked lime is mixed with sugar in different ratio (commonly 1:2) it shows binding properties like Plaster of Paris. After mixing with sugar, the semi-liquid paste dries quickly (four to five minutes), hardens, and binds the surfaces of interest. A cotton cloth was used in this experiment as a supporting material on which a hard surface can be formed. Slaked lime was spread over the cloth and then sugar was added onto the surface to make a quick paste and thus within four to five minutes a hard structure was formed similar to Plaster of Parris (Figure 15). Figure 16 shows two blocks made up with sugar and CaO after adding water. The blocks are found to have a hard structure with sufficient compressive strength. This mechanism can be applied for other purposes, i.e., to make insulation board, slabs, mold etc.

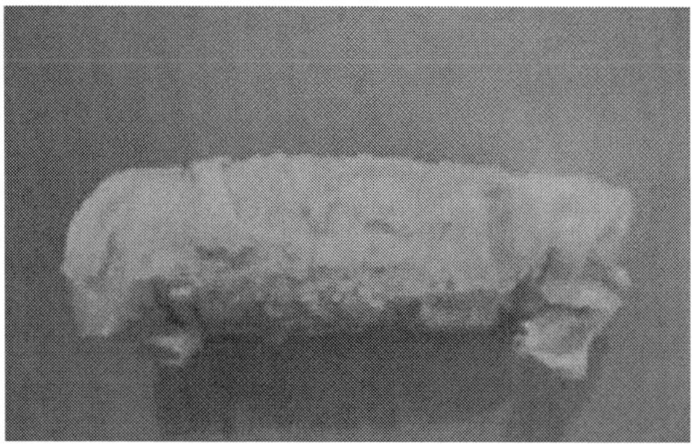

Figure 15. Alternative to Plaster of Paris.

4.7.4.6. Making of Particle Boards

Honey, CaO and sawdust were mixed in the ratio of 2:1:4 to make particle board as shown in Figure 17. It took approximately 16 hours to dry completely. The strength of the particle board was found satisfactory. It was strong enough to resist rupture upon several strikes of hammering.

Figure 16. Blocks made up with sugar with CaO.

Figure 17. Particle board made up with honey and CaO as adhesive.

4.7.4.7. Making of Sand Blocks

CaO and icing sugar were also used with sand and water to form a sand block. In this study, CaO, icing sugar and sand at a ratio of 1:2:5 mixed with sufficient amount of water to make a paste of a sand block (Figure 18) It was dried within a hour and a strong sand block was found.

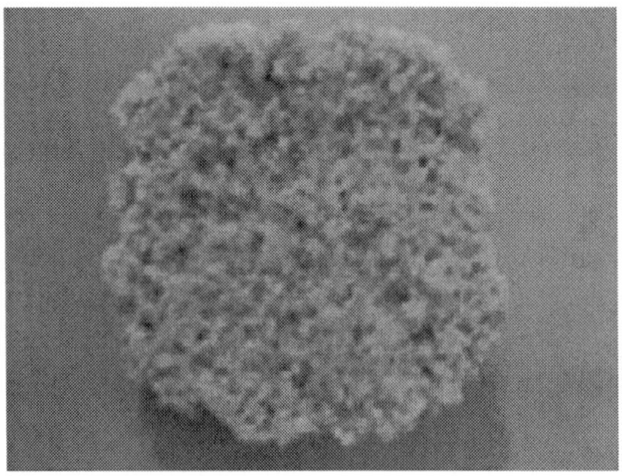

Figure 18. Sand block made up with sands and adhesive made up with sugar and CaO.

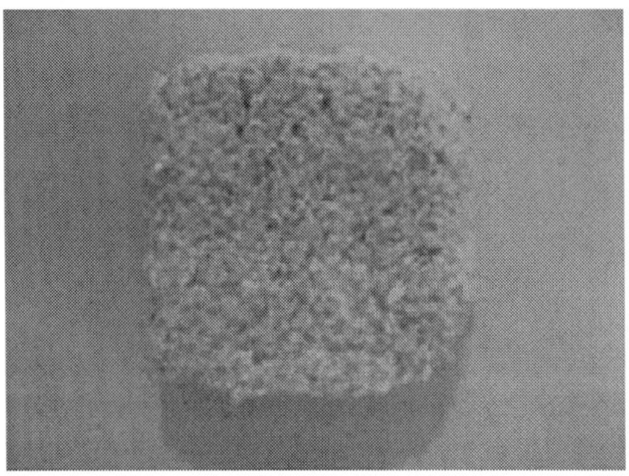

Figure 19. Sand block made up with sands and adhesive made up with honey and CaO.

This is sparingly water soluble and can be use for making mold, slab and interior surface. The same experiment was performed with adding honey and CaO with sands. Figure 19 shows a sand block which was made up with honey, CaO and sands with a ratio of 6:1:10. It took about two hours to dry this block.

The strength was adequate for daily handlings, with resistance to abrasion. The experiments show that these blocks can be used for interior design, mold or insulated board with higher strength.

CONCLUSIONS

Living in healthy environments is very important for both short- and long-terms. The use of natural adhesives made through sustainable technology is one of the steps towards the healthy environment. Nature is the best solution for all sustainable technologies. Therefore, Nature offers the key to sustainable technology developments. This paper has shown the effectiveness of a number of adhesive materials obtained from natural sources. They either meet or excel the standards set forth for commercial applications that have rarely seen truly natural products. This demonstrates the effectiveness of these products to handle short-term needs while assuring long-term environmental sustainability. In order to expand the application domain of these adhesives, more research should be planned. This line of products have the potential of replacing synthetic materials that have been implicated in the current environmental crisis, termed as 'technological disaster' by the Nobel Laureate (in Chemistry), Robert Curl.

ACKNOWLEDGMENT

The authors would like to acknowledge the contribution of the Atlantic Innovation Fund (AIF) and NSERC.

REFERENCES

AFP (Agence France Presse), (2007) *Honey Could Save Diabetics from Amputation*, News release, May 4.

Baumann, M. G. D., Lorenz, L.F., Batterman, S.A. and Zhang, G., (2000) Aldehyde Emissions from Particleboard and Medium Density Fiberboard Products, *Forest Products Journal*, vol. 50, No. 9, pp. 75-82.

Chhetri, A.B. and Islam, M.R., (2007a) Towards Producing Truly Green Biodiesel, *Energy Sources*, Accepted.

Chhetri, A.B. and Islam, M.R., (2007b) *Developing Inherently Sustainable Technologies*, Nova Science Publishers, New York, in press

CPSC, (2005) An Update on Formaldehyde: 1997 Revision [online] Available: (http://www.cpsc.gov/CPSCPUB/PUBS/725.html) [February 6, 2005]

Dhara, S. and Bhargava, P., (2001) Egg White as an Environmentally Friendly Low-Cost Binder for Gelcasting of Ceramics, *Journal of the American Ceramic Society*, vol. 84, No.12, pp. 3048-3050.

Elias, J. D. and Kraut, A., (1996) Formaldehyde Levels in Community Air Near a Particle Board Plant, *Proceedings of the Air and Waste Management Association's Annual Meeting and Exhibition*, 96-MP10.01, p.12.

Johansson, I., Karlsson, T. and Wimmerstedt, R., (2004) Volatile Organic Compound Emissions when Drying Wood Particles at High Dewpoints, *Chinese Journal of Chemical Engineering,* vol. 12, No. 6, pp. 767-772.

Khan, M.I. and Islam, M.R., (2007) True Sustainability in Technological Development and Natural Resources Management. Nova Science Publishers, New York, USA, 381 pp.

Kiosseoglou, V. (2003) Egg Yolk Protein Gels and Emulsions, Current Opinion in Colloid and Interface. *Science,* vol. 8, No. 4-5, pp 365-370.

LaFleur, L., Tatum, V.L., Jain, A. and Someshwar, A., (2004) Canadian Environmental Protection Act Aldehydes (Formaldehyde, Acetaldehyde, and Acrolein) and the Forest Products Industry, *NCASI Technical Bulletin*, No. 874, p.27.

Lambuth, A. L., (1977) Adhesives in the Plywood Industry, *Adhesives Age*, vol. 28, No. 4, pp. 21-26.

Miralai, S., (2006) *Replacing Artificial Additives with Natural Alternatives*, MASc Thesis, Dalhousie University, Nova Scotia, Canada.

Petrie, E. M., (2000) *Handbook of Adhesives and Sealants*, McGraw-Hill, Inc., NY.

Popa, M. and Popa, M.S., (2002) The Use of Volatile Organic Compounds as an Indicator in Indoor Air Quality Investigations, *Proceedings of SPIE - The International Society for Optical Engineering*, vol. 4900, No. 2, pp 1090-1093.

Prane, J. W., (1996) Newly Revised Rauch Guide Remains Comprehensive Source, Adhesives Age, originally from "*The Rauch Guide to the U.S. Adhesives and Sealants Industry, 1995–1996 Edition*", Impact Marketing Consultants, Inc., Manchester Center, VT.

Rathna, G.V.N., Li, J. and Gunasekaran, S. (2004) Functionally-Modified Egg White Albumen Hydrogels, *Polymer International*, vol. 53, No. 12, pp. 1944-2000.

Rogers, I., Mahood, H., Servizi,J. and Gordon, R., (1979) Identifying Extractives Toxic to Aquatic Life, *Pulp and Paper Canada*, vol. 80, No. 9, pp 94-96, 98-99.

Sabater, M.C., Martinez, M.A. and Font, R., (2001) Toxicity and Hazardous Properties of Solvent Base Adhesive Wastes, *Waste Management and Research*, vol.19, No. 5, pp. 442-449.

Shin, H. S. and Ustunol, Z., (2005) Carbohydrate Composition of Honey from Different Floral Sources and their Influence on Growth of Selected Intestinal Bacteria: An In Vitro Comparison, *Food Research International*, vol. 38, No. 6, pp. 721-728.

Steenhuysen, J., (2007) Mother Nature Still a Rich Source of New Drugs , Environmental News Networks [online] Available: (http://www.enn.com/med.html?id=1442) [April 15, 2007].

Wilson, B. (2004) *The Hive: The Story Of The Honeybee*, London, Great Britain: John Murray (Publishers).

Yakovskaya, M.E., (1993) Industrial Hygiene in Plywood Production, *Meditsinskaya Radiologiya I Radiatsionnaya Bezopasnost*, n 11-12, Nov-Dec, pp. 20-22.

Zatzman, G.M., (2007) The Honey → Sugar → Saccharin® → Aspartame®, or HSS®A® Syndrome: A note, *J. Nature Science and Sustainable Technology*, vol. 1, No. 3.

Zatzman, G.M. and Islam, M.R., (2007) Truth, Consequences and Intentions: The Study of Natural and Anti-Natural Starting Points and their Implications, *J. Nature Science and Sustainable Technology*, vol. 1, No. 2, pp. 1-38.

INDEX

A

B

C

D

E

F

G

H

I

J

K

L

M

N

O

P

W

X

Y